DIETARY FIBER FOR THE PREVENTION OF CARDIOVASCULAR DISEASE

DIETARY FIBER FOR THE PREVENTION OF CARDIOVASCULAR DISEASE

Fiber's Interaction Between Gut Microflora, Sugar Metabolism, Weight Control and Cardiovascular Health

Edited by

RODNEY A. SAMAAN, MD, MPH
ProgessiveMD
Los Angeles, CA, United States

Academic Press is an imprint of Elsevier
125 London Wall, London EC2Y 5AS, United Kingdom
525 B Street, Suite 1800, San Diego, CA 92101-4495, United States
50 Hampshire Street, 5th Floor, Cambridge, MA 02139, United States
The Boulevard, Langford Lane, Kidlington, Oxford OX5 1GB, United Kingdom

Notices

Knowledge and best practice in this field are constantly changing. As new research and experience broaden our understanding, changes in research methods, professional practices, or medical treatment may become necessary.

Practitioners and researchers must always rely on their own experience and knowledge in evaluating and using any information, methods, compounds, or experiments described herein. In using such information or methods they should be mindful of their own safety and the safety of others, including parties for whom they have a professional responsibility.

To the fullest extent of the law, neither the Publisher nor the authors, contributors, or editors, assume any liability for any injury and/or damage to persons or property as a matter of products liability, negligence or otherwise, or from any use or operation of any methods, products, instructions, or ideas contained in the material herein.

Library of Congress Cataloging-in-Publication Data
A catalog record for this book is available from the Library of Congress

British Library Cataloguing-in-Publication Data
A catalogue record for this book is available from the British Library

ISBN: 978-0-12-805130-6

For information on all Academic Press publications visit our website at https://www.elsevier.com/books-and-journals

Publisher: Mica Haley
Acquisition Editor: Stacy Masucci
Editorial Project Manager: Sam Young
Production Project Manager: Lucía Pérez
Designer: Matthew Limbert

Typeset by Thomson Digital

CONTENTS

CONTRIBUTORS

Vanesa Benítez
University Autonoma, Madrid, Spain

Domenico Bonaduce
University of Naples "Federico II", Naples, Italy

Antonio Calignano
University of Naples "Federico II", Naples, Italy

Roberto B. Canani
CEINGE Advanced Biotechnologies, University of Naples "Federico II", Naples, Italy

Anil K. Chauhan
Center of Food Science and Technology, BHU, Varanasi, Uttar Pradesh, India

Carmen di Scala
University of Naples "Federico II", Naples, Italy

Rosa M. Esteban
University Autonoma, Madrid, Spain

David Feder
Federal University of ABC, Faculty of Medicine, Santo André, São Paulo, Brazil

Fernando L.A. Fonseca
Federal University of ABC, Faculty of Medicine, Santo André, São Paulo, Brazil

Ashish Gurav
University of Pennsylvania School of Dental Medicine, Philadelphia, PA, United States

Santosh K. Jha
Department of Bioengineering Birla Institute of Technology, Mesra, Jharkhand, India

Pamela M. Martin
Medical College of Georgia, Augusta University, Augusta, GA, United States

Esperanza Mollá
University Autonoma, Madrid, Spain

Deepak Mudgil
Mansinhbhai Institute of Dairy & Food Technology, Mehsana, Gujarat, India

Mohammad A. Niaz
Halberg Hospital and Research Institute, Moradabad, Uttar Pradesh, India

Lorella Paparo
University of Naples "Federico II", Naples, Italy

Pragya Prakash
Department of Bioengineering Birla Institute of Technology, Mesra, Jharkhand, India

Puttur D. Prasad
Medical College of Georgia, Augusta University, Augusta, GA, United States

Divya R. Gunashekar
University of Minnesota—School of Public Health, Minneapolis, MN, United States

Roberto Russo
University of Naples "Federico II", Naples, Italy

Rodney A. Samaan
ProgessiveMD, Los Angeles, CA, United States

Anand R. Shewale
University of Arkansas for Medical Sciences, Little Rock, AR, United States

Hare R. Singh
Department of Bioengineering Birla Institute of Technology, Mesra, Jharkhand, India

Nagendra Singh
Medical College of Georgia; Cancer Research Center, Augusta University, Augusta, GA, USA

Ram B. Singh
Halberg Hospital and Research Institute, Moradabad, Uttar Pradesh, India

Ravi P. Singh
Institute of Agricultural Sciences, BHU, Varanasi, Uttar Pradesh, India

Toru Takahashi
Fukuoka University, Fukuoka, Japan

Carlo G. Tocchetti
University of Naples "Federico II", Naples, Italy

Matam Vijay-Kumar
Pennsylvania State University Medical Center, Hershey, PA, United States

Huabin Zhu
Medical College of Georgia, Augusta University, Augusta, GA, United States

CHAPTER 1

Dietary Fiber and Human Health: An Introduction

Santosh K. Jha, Hare R. Singh, Pragya Prakash

1 INTRODUCTION

Fibers are the structural part of a plant and are found in all plant foods, including vegetables, fruits, grains, and legumes. Dietary fiber is a discrete group of carbohydrate found almost exclusively in plants, including non-starch polysaccharides (NSPs), such as cellulose, pectin, and lignin. It is the edible integral element of carbohydrate and lignin that is naturally found in plant food. Most dietary fibers are polysaccharides are starches, but they are not digested by humans. Starch is a long chain of glucose molecules linked together with alpha bonds. Fiber is a long chain of glucose molecules linked together with beta bonds. The human body lacks enzymes to break beta bonds, therefore fiber is not digested and absorbed. The undigested fiber passes into the lower intestine where intestinal bacteria can ferment the fibers.

Fibers are typically known as NSPs. NSP fibers include cellulose, hemicellulose, pectin, gums, and mucilages. Dietary fibers act by changing how other nutrients and chemicals are absorbed. Some varieties of soluble fiber absorb water to become a gelatinous, viscous substance that is degraded by microbes in the digestive tract. Some types of insoluble fiber have bulking action and are not fermented. Lignin, a significant dietary insoluble fiber may alter the rate and metabolism of soluble fibers. Some other types of insoluble fiber, notably resistant starch (RS), are fully fermented. Some but not all soluble plant fibers block intestinal mucosal adherence and translocation of potentially pathogenic bacteria and may therefore modulate intestinal inflammation, an effect that has been termed contrabiotic.

The occurrence of the fiber depends on the sources and origin. For example, in vegetables, fiber is mainly concentrated on the skin, but trace amounts are found in the rind too, while in cereals it is mainly found in the bran portion, whereas in fruits it is not only present in the skin but

Dietary Fiber for the Prevention of Cardiovascular Disease
http://dx.doi.org/10.1016/B978-0-12-805130-6.00001-X

Table 1.1 A classification of the unavailable carbohydrates in foods [1]

Major sources in the diet	Description	Classical nomenclature
Structural materials of the plant cell wall	Structural polysaccharides Noncarbohydrate constituents	Cellulose, some pectins, and hemicellulose Lignin, minor constituents
Nonstructural materials, either found naturally or used as food additives	Polysaccharides from a variety of sources	Pectins, gums, mucilages, algal polysaccharides, chemically modified polysaccharides

also in the fruity portion named the mesocarp. Dietary fiber usually refers to the unavailable carbohydrates, and is mainly classified as soluble and insoluble. The insoluble fibers are present in fruits, vegetables, grains, legumes, and cereals. The soluble fibers include pectin, gums, certain hemicelluloses, and storage polysaccharides. Fruits, some vegetables, oat, barley, soybeans, psyllium seeds, and legumes contain more soluble fiber than other foods.

The American Association of Cereal Chemists (AACC, 2001) defined dietary fiber as: "the edible parts of plants or analogous carbohydrates that are resistant to digestion & absorption in the human small intestine with complete or partial fermentation in the large intestine. Dietary fiber includes polysaccharides, oligosaccharides, lignin, and associated plant substances. Dietary fiber promotes beneficial physiological effects including laxation, blood cholesterol attenuation, and blood glucose attenuation."

These unavailable carbohydrates can be categorized into two major groups, based on the structural components as shown in Table 1.1.

2 TYPES OF DIETARY FIBER

Dietary fiber comprises cellulose, noncellulosic polysaccharides (NCPs), such as hemicelluloses, pectic substances, and a noncarbohydrate component lignin. These are mainly the structural components of the plant cell wall. Some types of plant fiber are not cell wall components, but are formed in specialized secretory plant cells including plant gums. Gums are sticky exudations formed in response to trauma. Mucilages are secreted into the endosperm of plant seeds where they act to prevent excessive dehydration and include materials with widespread industrial applications, such as guar gum.

2.1 Soluble Versus Insoluble Fiber

Chemists classify fibers depending on how promptly they dissolve in water, so these can be categorized as soluble fibers and insoluble fibers. The insoluble fibers act as a sponge within the gut by absorbing water. This will increase the softness and bulk of the stool and will thereby decrease the chance of constipation, and colon carcinoma [2]. Scientists have conjointly recommended that soluble fibers can reduce glucose levels and hence prevent diabetes [3]. Foods containing fiber often have a combination of both soluble and insoluble fibers. Diets rich in soluble fiber includes dried beans and oranges. Insoluble fibers include whole wheat (wheat bran) and rye.

The Institute of Medicine recommends that the terms soluble and insoluble fibers should no longer be used to classify dietary fibers even though they may still appear on some food labels. Both are considered to be functional fibers; nondigestable carbohydrates have beneficial physiological effects in humans. Another new term is total fiber, which is the sum of dietary fiber and functional fiber [4].

2.1.1 Common Fibers

The most common fibers are cellulose, hemicellulose, and pectic substances. A few other types of fiber include vegetable gums, inulin, beta-glucan, oligosaccharides, fructans, some RSs, and lignin. Lignin belongs to the class of the fibers that are not carbohydrates [4].

2.2 Polysaccharides

Cellulose, the foremost verdant molecule found in nature, is the beta compound of starch; it is a long (up to 10,000 sugar residues) linear compound of 1,4 joined units of glucose. Hydrogen bonding between sugar residues in adjacent chains imparts a crystalline microfibril structure; cellulose is insoluble in robust alkali. NCPs embrace an oversized variety of heteroglycans that consist of a mixture of pentoses, hexoses, and uronic acids. Among a lot of vital NCPs are hemicellulose and cellulose substances [5]. Hemicelluloses are cell wall unit polysaccharides solubilized by a binary compound alkali after the removal of water soluble and cellulose polysaccharides (Table 1.1). The hemicelluloses are subclassified on the premise of the principal monomeric sugar residue. Acidic and neutral forms differ by the content of glucuronic and galacturonic acids. Uronic acid formation involves the reaction of the terminal —CH_2OH to —COOH, and is of biological importance since the sugar residues become obtainable for methylation, amidation, and formation of ion complexes [6]. Hemicelluloses, particularly the simple sugar and uronic acid parts, are accessible to microorganism enzymes as compared to cellulose [7,8]. Pectic substances are a cluster of polysaccharides within which D-galacturonic acid is a major constituent. They are the structural part of plant cell walls and additionally act as cementing material. These substances embrace a water-insoluble parent compound protopectin, further as pectinic acids and pectin. Long chains of galacturonan are interrupted by blocks of L-rhamnose-rich units that end up in bends within the molecule. Several pectins have neutral sugars covalently joined to them as side chains, chiefly arabinose and sucrose, sugar, rhamnose, and aldohexose. It has additionally been observed that small quantities of glucuronic acid are also joined to cellulose in a facet chain [9–11]. The carboxyl groups of the galacturonic acids are partly methylated, and therefore the secondary hydroxyls are also acetylated. Pectin is extremely soluble and is metabolized by colonic bacterium [12].

2.3 Lignin

Lignin is not a sugar, but is rather a compound containing about 40 oxygenated phenylpropane units, as well as coniferyl, sinapyl, and *p*-coumaryl alcohols that have undergone a dehydrogenative polymerization process [13,14]. Lignins vary in mass and methoxyl content. As a result of strong intramolecular bonding which has carbon-to-carbon linkages, lignin is inert. Lignin displays a greater resistance to digestion than the other naturally occurring compound.

Lignin is a fiber that is not sugar, but rather a saccharide, consisting of long chains of phenolic resin alcohols connected along an oversized advanced molecule. As plants mature, their cell walls increase in lignin concentration, leading to a tough, stringy texture. This partly explains why celery and carrots get harder as they age. Boiling water does not dissolve or even soften the lignin.

3 PHYSIOLOGICAL EFFECTS OF DIETARY FIBERS

See Fig. 1.1.

3.1 Dietary Fibers Effects on Immunity

The epithelial duct is subjected to monumental and continuous foreign substance stimuli from food and microbes; it will integrate complicated interactions among diet, external pathogens, and native immunologic and

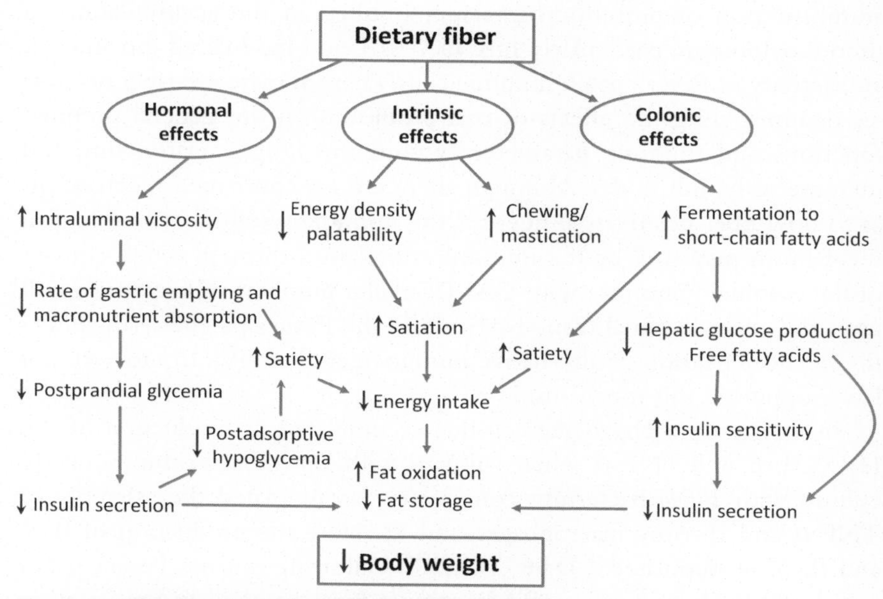

Figure 1.1 ***Physiological effects of dietary fibers [15].***

nonimmunological processes. There is an increasing proof that fermentable dietary fibers and the new delineated probiotics could modulate numerous properties of the system, as well as those of the gut-associated lymphoid tissues (GALT).

3.1.1 Proposed Mechanisms: Immunomodulating Effects of Dietary Fibers

- Direct contact of lactic acid bacteria or bacterial products (cell wall or cytoplasmic components) with immune cells in the intestine.
- Production of short-chain fatty acids from fiber fermentation.
- Modulation of mucin production.

3.1.1.1 Direct Contact of Lactic Acid Bacteria or Bacterial Products (Cell Wall or Cytoplasmic Components) With Immune Cells in the Intestine

It is typically assumed that the consumption of prebiotics, through their effects on the colonic microflora, can have an analogous impact as probiotics on the performance of the immune system. Studies have documented the effects of feeding carboxylic acid bacterium (i.e., *Lactobacilli* and *Bifido* bacteria) on varied parameters of immune function. Oral administration of probiotic bacterium resulted in the accumulation of immunoglobulins, particularly IgA, in GALT and modulated the amount and activity of Peyer's patch immune cells. There also are a variety of studies demonstrating the effects of oral probiotics on the general immune functions and immune parameters within the lungs, peritoneum, and mesenteric lymph nodes. Although there are no confirmed mechanisms as to how fiber positively influences the immune system, but one logical mechanism may well be the immune stimulation through direct contact of the colonic microflora with GALT. Smaller numbers of bacterium will cross the enteral animal tissue barrier into the Peyer's patches [16] resulting in the activation of alternative immune cells [16–19]. In vitro studies have supported this mechanism.

In one study, a phagocyte cell line accumulated its production of gas, H_2O_2, IL-6, and TNF-α when cultured with *Bifido* bacterium. Similarly, culture with *Bifido* bacterium considerably accumulated the assembly of TNF-α and IL-6 by macrophages, and therefore the production of IL-2 and IL-5 by stimulated CD4+ cells [20]. Culturing murine Peyer's patch cells with *Bifido* bacterium (*Bifidobacterium breve*) resulted in proliferation and protein production by B-lymphocytes and activated macrophage-like cells [21].

3.2 Susceptibility to Bacterial Enzyme Degradation

Fermentation could also exert a physiological impact or alter the chemical atmosphere of the cecum, thus affecting the growth or the metabolic activity of the microorganism.

The extent of fiber degradation within the colon relies on the character of the colonic microorganism flora; the condition of a given fiber to microorganism digestion relies on its physical and chemical structure. Digestion of polysaccharides varies between 30% and more than 90%. Pectin and hemicellulose are lost through the stool; cellulose is somewhat less well digestible [22–24]. Lignin, by virtue of its chemical compound cross-linked structure, is resistant against microorganism degradation and is nearly utterly recovered within the stool. Polysaccharides from extremely hard plant tissue areas are less well digestible because physical encrustation and chemical bonding to lignin occur.

3.3 WaterRetention Capability of Dietary Fiber

The water retention capability of dietary fiber has necessary physiological effects in each of the small and large intestines.

The initial event upon exposure of fiber to an aqueous medium is surface adsorption of water molecules. Particle size may additionally influence the water retention capability of fiber [25–27]. Robertson and Eastwood have demonstrated that the fiber preparation alters the water retention capabilities. This implies that the physical structure of fiber is the most significant determinant of hydratability.

3.4 Adsorption of Organic Materials

Adsorption of digestive fluid acids has been best documented and relies on the composition of the fiber, the chemistry of the steroid, and therefore the pH and osmolality of the encompassing medium [28,29]. Lignin is the most potent steroid adsorbent, and binding is seemingly influenced by mass, pH, and therefore the presence of methoxyl and α-carbonyl groups on the lignin molecule [30]. Eastwood and Hamilton [31] demonstrated that methylation of polymer accumulated steroid sorption. These authors conjointly reported that adsorption was greatest at low pH. Each condition would either block or suppress ionization of carboxyl groups and hydroxyl groups on lignin's phenyl propane units, suggesting a hydrophobic bonding mechanism. These studies predict that interaction with lignin is greater for bacterially changed digestive fluid acids shaped within the colon. Saponins conjointly bind bile acids in vitro [32] and in vivo [33], and they may be responsible for sterol adsorption related to fiber. However, bran and alfalfa showed no diminution of adsorbent capability following removal of associated saponins [34].

3.5 Cation Exchange Properties

The useful capability of dietary fiber for ion exchange is well established. The impact is expounded to the amount of free carboxyl teams on the sugar residues [35,36]. Calcium binding is foretold on the premise of uronic acid content of fiber residues [37]. Formation of ion complexes with acidic polysaccharides is mirrored in their effects on mineral balance, solution absorption, and significant metal toxicity.

3.5.1 Quantifying Fiber Content

Several completely different laboratory strategies are used to measure the quantity of fiber in foods. The older technique consisted of treating a food with robust acid to simulate the surroundings of the abdomen, so treating it with a base to parallel the experience within the small intestine. The remaining weight of undigested fiber was measured as "crude fiber" and was listed in most food composition tables as "fiber." This rather inaccurate technique has been for the most part replaced by associate analytical technique approved by the Association of Official Analytical Chemists International (AOAC) that measures dietary fiber. For each 1 g of crude fiber, there are 2–3 g of dietary fiber.

4 DIETARY FIBER SUBSTANCES

4.1 Inulin

This polysaccharide consists of continuous repetitive units of fruit sugar with associate finish molecule of glucose. Although this fiber happens naturally in over 30,000 plants, it is most typically found in asparagus, artichoke, and garlic. Food processors extract polysaccharide from the chicory root [31], and this soluble fiber is used by the industry to impart a creamy texture to frozen farm merchandise like no-fat or no-sugar frozen dessert, improve the textures of spread spreads, and develop no-fat icings, fillings, and whipped toppings (Table 1.2).

5 RECOMMENDED FIBER INTAKE

Fiber is a carbohydrate that is not digested or absorbed, so it contributes few, if any, calories. The Institute of Medicine recommends that healthy adults consume:

- 25 g of daily fiber for women
- 38 g of daily fiber for men

Fiber recommendations for children and elderly are 14 g of fiber for every 1000 calories (kcal) consumed. The average US intake of only 15 g of dietary fiber per day falls short of these goals. One of the ways to increase

Table 1.2 Dietary fibers and their prebiotic effects

Fibers	Prebiotic effects
Wheat dextrin	Increase bacteroides, reduce clostridium
Inulin	Bifidogenic
Galactooligosaccharide	Bifidogenic
Acacia gum	Bifidogenic
Psyllium	Prebiotic potential
Polydextrose	Bifidogenic
Whole grain	Prebiotic potential
Banana	Fecal microbiota

both complex carbohydrates and fiber is by offering a vegetarian menu option.

Whole grain cereals, fruit and vegetables, and potatoes comprise most sources of dietary fiber. Cellulose occurs along with hemicelluloses in cereals. The hard outer layers are the predominant fiber supply in whole-grain product. Oats and barley contain high concentrations of a soluble, viscous sort of sugar, β-glucan. Pectins, a main source of dietary fiber in fruits and vegetables, have similar properties (Tables 1.3 and 1.4).

5.1 Total Carbohydrates

Average sugar intake in adolescents in European countries varied between 41 and 58 g. Mean intakes were highest within the Czech Republic and Norway and lowest in Greece and Spain.

5.2 Summary of Dietary Reference Values and Suggestions

A variety of national and international organizations have set dietary reference values (DRVs) for carbohydrates (total and/or glycemic) and also for dietary fiber intake. Generally, reference intake are expressed as percentage of the overall energy intake (%). For fiber, intakes are expressed in grams per day and/or on an energy basis (per MJ or per 1000 kcal).

5.2.1 Glycemic Carbohydrates

As per the Nordic Nutrition Recommendations of 2004 [40], carbohydrates (including energy from dietary fiber, 8 kJ/g) ought to offer 50%–60%

Table 1.3 Recommended levels of fiber intake [38]

S. no.	Sources	Year	Recommendations
1	USDA and USFDA	1993	TDF 25 g/day
2	Federation of American Society of Experimental Biology	1987	TDF 20–25 g/day; IDF 70%–75%; SDF 25%–30%
3	American Dietetic Association	1988	TDF 20–35 g/day
4	Dutch RDA (Voedingsraad)	1989	12.5 g/1000 kcal/day
5	British National Advisory	1986	TDF 30 g/day
6	UK Department of Health	1991	NSP 18 g/day
7	German RDA	1991	2.5 g/kcal/day
8	Expert Advisory Committee on dietary fiber—Health and Welfare Canada	1985	5.8–8.0 g/1000 kcal/day

IDF, Insoluble dietary fiber; SDF, soluble dietary fiber; TDF, total dietary fiber.

Table 1.4 Dietary fiber profile of foods [39]

Food source	Serving size	Total fiber (g)	Soluble fiber (g)	Insoluble fiber (g)
Breads, cereals, and pasta				
Cornflakes	1 cup	0.5	0.0	0.5
White bread	1 slice	0.53	0.03	0.5
Whole grain bread	1 slice	2.9	0.08	2.8
White rice	1/2 cup cooked	0. 5	0.5	0.0
Brown rice	1/2 cup cooked	1.3	1.3	0.0
Spaghetti	1/2 cup cooked	0.8	0.02	0.8
Bran (100%) cereal	1/2 cup	10.0	0.3	9.7
Rolled oats	3/4 cup cooked	3.0	1.3	1.7
Oats, whole	1/2 cup cooked	1.6	0.5	1.1
Corn grits	1/2 cup cooked	1.9	0.61	0.3
Popcorn	3 cups	2.8	0.8	2.0
Fruits				
Apple	1 small	3.9	2.3	1.6
Apricots	2 medium	1.3	0.9	0.4
Banana	1 small	1.3	0.6	0.7
Grapefruit	1/2 fruit	1.3	0.90	0.4

of the overall energy intake. The population goal is 55% from carbohydrates that ought to be used for designing purposes. The intake of refined, supplemental sugars should not exceed 10%. The Health Council of Netherlands [41] supported their recommendations for light carbohydrates [42]. A sugar intake above 55 seems to be related to risk of dyslipidemia (e.g., exaggerated VLDL and small HDL-cholesterol).

The German–Austrian–Swiss recommendations are discussed in Ref. [43]. It is stressed that carbohydrates ought to be derived from foods rich in starch and dietary fiber, and intake of refined sugars ought to be restricted. In order to avoid gluconeogenesis from macromolecule (e.g., amino acids) and to inhibit lipolysis, a minimum of 25% of the energy ought to be provided from carbohydrates. This proportion applies to all ages. No quantitative recommendation was created for (added) sugars.

WHO offers population nutrient intake goals (population average intakes that area unit judged to be in step with the upkeep of health in a very population) for preventing diet-related chronic diseases [44]. For carbohydrates the population goal is ready at 55–75%, together with dietary fiber. A recent FAO/WHO Scientific Update on saccharide within the human diet proposes that the variation is extended to 50–75% [45]. The

UK Committee on Medical Aspects of Food Policy [46] set a dietary reference (population average intake) for starches and intrinsic and milk sugars of 37%. For nonmilk sugars the population average intake must not exceed 60 g/day, primarily based on the role of frequent consumption of such sugars. The US Food and Nutrition Board calculated the typical demand of (glycemic) carbohydrates as 100 g/day for youngsters and adolescents up to 18 years, further as adults [4], primarily based chiefly on information relating to glucose employment by the brain. The US Food and Nutrition Board additionally set Acceptable Macronutrient Distribution Ranges (AMDR) for total carbohydrates of 45–65% for individuals. The AMDR are supported by a proof indicating an attenuated risk for coronary heart disease (coronary heart disease, CHD) at low intakes of fat and high intakes of carbohydrates.

The Health Council of Netherlands [42] has set tips for the intake of dietary fiber which chiefly supported the importance of dietary fiber for intestinal functioning and its relationship to risk of coronary heart disease. For youngsters a gradual increase in the intake is suggested. It increases as the youngsters move up in age group. The suggested consumption is 2.8 g/MJ for 1–3 years, 3.0 g/MJ for 4–8 years, 3.2 g/MJ for 9–13 years, and 3.4 g/MJ from 14 years of age, respectively. Within the Nordic Nutrition Recommendations, the intake limit of dietary fiber is set as 25–35 g/day in adults, that is, 3 g/MJ [40]. Adequate intake of dietary fiber reduces the danger of constipation and might presumably contribute to protection against carcinoma. No recommendation is given for youngsters as a result of restricted evidence; however, it is declared that intake of acceptable amounts of dietary fiber from a spread of foods is vital for youngsters. Within the UN agency report of 2003 [44], there's no precise population goal for the intake of total dietary fiber, rather a minimum of 25 g/day ought to be provided from fruit, vegetables, and whole-grain foods. This population goal relies on evidence linking high intake of dietary fiber (from fruit, vegetables, and whole-grain foods) with attenuated risk of weight gain, diabetes type 2, and cardiovascular diseases (CVDs). The food-based recommendation was supported by the recent FAO/WHO scientific update on saccharide within the human diet. Within the biological process recommendations for the French population, intake of dietary fiber of greater than 25 g/day is suggested for maintaining "a healthy colon" and to decrease the danger of carcinoma, with 30 g/day as a most popular level. Also, increased intake of dietary fiber is declared to be advantageous in conditions like dyslipidemia and in diabetes mellitus type 2. The guiding value for dietary fiber within the German-Austrian-Swiss recommendations is a minimum of 30 g/day, similar to 3 g/MJ for women and 2.4 g/MJ for men, respectively [43]. The basis of these values is studies associating exaggerated dietary fiber intake with a decreased risk of constipation, carcinoma, obesity, hypercholesterolemia, diabetes mellitus type 2, and coronary artery disease. The United Kingdom Committee on Medical Aspects of Food Policy [46] set a dietary reference for NSP to 18 g/day, with a personal variation of 12–24 g/day. The US Food and Nutrition Board [4] set an adequate intake for total dietary fiber of 3.4 g/MJ (14 g/1000 kcal) which supported the energy-adjusted median intake related to all-time low risk of CHD in empiric studies. The adequate intake is applicable for all age and sex classes from 1 year and older. The adequate intake corresponds to 25 g/day for women and 38 g/day for men aged 14–50 years, respectively. According to European Society for Pediatric Gastroenerology Hepatology and Nutrition (ESPGHAN) individuals of college age require a diet which will be probably going to supply a minimum of 10 g/day of dietary fiber [47].

5.3 Criteria (Endpoints) for Determining Dietary Reference Values

The number of carbohydrates and dietary fiber within the diet could have an effect on each short-run and long metabolic responses like liquid body substance lipids or plasma glucose and internal secretion concentrations, which may be thought to be criteria for establishing dietary reference values (DRVs). LDL-cholesterol has been causally associated with the chance of developing vascular diseases [4], whereas triglycerides, LDL/HDL quantitative relation, or total cholesterol/HDL quantitative relation have additionally been related to disorder (CVD) risk in epidemiologic studies [48]. In addition to CVDs, alternative long endpoints for establishing DRVs for each glycemic carbohydrates and dietary fiber embody weight management, proper functioning of intestine, polygenic disorder, and a few cancers. The DRVs apply to healthy populations and that they are not supposed as reference values for the treatment of patients with diseases or conditions, such as polygenic disorder, obesity, or CVD. However, they apply to healthy subjects with signs of metabolic disturbances, such as impaired glucose tolerance, elevated blood pressure, or serum lipids. Aside from carbohydrates and dietary fiber, the number and sort of fat and protein within the diet additionally influence these metabolic factors.

6 DIETARY NECESSITIES AND FUNCTIONING

Dietary fiber has not been shown to be an imperative part of the diet. However, dietary fiber incorporates a major role in bowel functions and gastrointestinal symptoms, such as constipation [4].

6.1 Gastrointestinal Function

6.1.1 Fiber in Adults

Dietary fiber plays a major role in bowel functions, whereas low fiber intakes leads to bowel dysfunction, such as constipation [47]. Constipation has been outlined as a problem in passing stools [48]. Constipation occurs in 5%–18% of adults in several countries, with a higher percentage of women affected,and adversely impacts the standard of life [49] (Fig. 1.2).

Constipation may conjointly contribute to diverticular disease. Each empirical and experimental knowledge shows that dietary fiber is the most significant dietary determinant of fecal bulk and transit time [51,52]. Dietary fiber from cereals, fruits, and vegetables could increase stool weight that promotes traditional laxation in youngsters and adults. In general, the bigger the burden of the stool and faster the speed of passage through the colon the higher the laxative result [45]. It has conjointly been demonstrated that completely different types of dietary fiber have different bulking capability. Dietary fiber in wheat bran and different fiber that is fairly resistant against fermentation within the giant intestine have the foremost pronounced bulking result (5–6 g/g dietary fiber) primarily attributable to water binding within the distal intestine. Although there is no single accepted definition of what constitutes a normal frequency of defecation, it is often regarded as once per day on western diets [53]. Haack et al. [54] have indicated that a frequency of once per day and a transit time within the 2–3 days could also

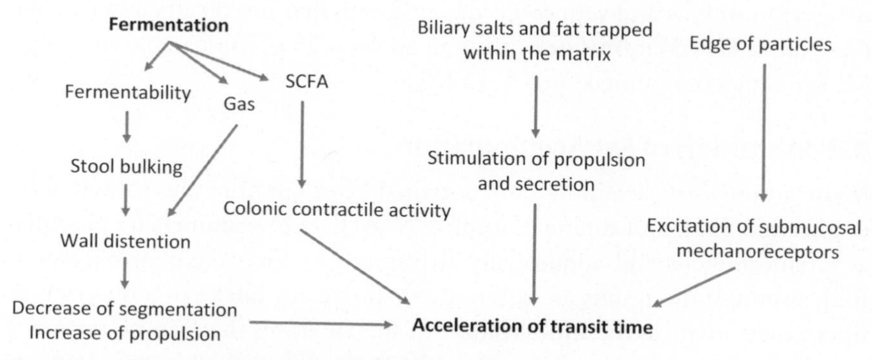

Figure 1.2 ***Gastrointestinal effect of dietary fibers [50].***

be thought of "normal". They reported that increasing intake of fiber (provided by a combination of fruit, vegetables, and grains) from 16 to 30 g/day enlarged defecation frequency from 0.7 to 0.94 times per day; an additional increase in fiber intake to 42 g/day had no further result on frequency; it remained at once per day. There was no important amendment over the variation of fiber intakes in gastrointestinal transit time that remained within 2–3 days. Fecal weight enlarged from 109 to 156 g and 195 g at fiber intakes of 16, 30, and 42 g/day, respectively. Increasing dietary fiber intake from 12 to 45 g/day enlarged fecal weight from 69 to 184 g/day and reduced transit time from 70 to 45 h [55]. Mean transit time in UK adults has been reported as 70 h (median 60 h) in a population within which mean stool weight is about 110 g/day, but 100 g/day in about 50% of people and dietary fiber intake of about 18 g/day [51]. It has been calculated that a dietary fiber intake of 25 g/day is related to stool weight of 130–150 g/day [56]. It was rumored that adults consumed 25 g dietary fiber in their usual diet excreted over a 150 g fecal matter per day [52]. Further these data indicate that an intake of 25 g/day of dietary fiber from mixed foods (as AOAC fiber or equivalent) is compatible with internal organ transit time of about 2–3 days and a defecation frequency of once per day and a fecal moisture of greater than 70% may be thought of adequate for most adults (Fig. 1.3).

6.1.2 Fiber in Children

There is evidence that constipation could be more common throughout childhood [57] in which there is an inverse relationship with dietary fiber intake [58]. The fiber intake among 1-year-old German kids was already higher (3 g/MJ) and there are no reports of adverse effects associated with the fiber intake.

In conclusion, dietary fiber intake of 2 g/MJ ought to be adequate for defecation in kids and supported the evidence that the dietary fiber intake is adequate for traditional defecation in adults is 25 g, comparable to 2–3 g/MJ for daily energy intakes of 8–12 MJ.

6.2 Inhibition of Fat Accumulation

Many authors have examined the potential of a high fiber diet to switch fat oxidization [59], and different studies [60,61] have examined its potential as satiation agent and additionally an ingredient for weight management [62], although the results are still not conclusive. An intake of a diet rich in fiber could increase the mobilization of the fat stores that can be used as a result of a decrease in insulin secretion [63]. Studies to this point in humans

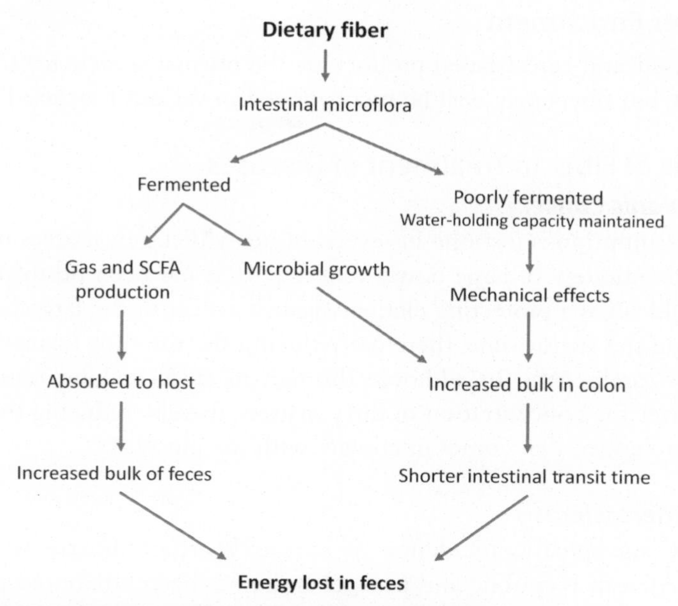

Figure 1.3 ***Effect of dietary fiber over intestinal microflora.***

would indicate that diets rich in fiber do not have an effect on total energy expenditure, saccharide oxidization, or fat oxidization [64–67]. In another study on human volunteers, breads rich in fiber imparted higher satiation than white breads taken between 70 and 120 min [68]. Higgins et al. examined the connection between the content of a meal and postprandial fat oxidization, finding that 5.4% of total dietary carbohydrates with fiber may considerably increase postprandial lipid oxidization and possibly cut back fat accumulation in the long run [69]. It had been reported that the employment of fiber within the diet as a bioactive purposeful food element could be a natural endogenous method to reduce energy intake, hence a natural approach to the treatment of obesity [70].

6.3 Absorption of Minerals

Fiber enhanced the ileal absorption of a variety of minerals in rats and humans. Fiber-rich diets increased absorption of calcium, magnesium, zinc, iron, and copper in rats [71,72]. Fiber may have a positive result on enteral calcium and iron absorption. A study to check the apparent enteral absorption of calcium, phosphorus, iron, and zinc within the presence of either resistant or light starch showed that a meal containing 16.4% fiber resulted in a bigger apparent absorption of calcium and iron compared with light starch [73].

6.4 Fiber Enrichment

Breads, food, and cereal-based product are the normal vehicle for fiber enrichment, but fibers may well be incorporated in various processed foods.

6.5 Role of Fiber in Treatment of Diseases

6.5.1 Colonic Cancer

Fiber consumption is just one in several of many dietary variables involved within the etiology of large bowel cancer [74]. It has been postulated that fiber could act as a protecting element against cancer of the large bowel by shortening the transit time, therefore reducing the time for formation and action of carcinogens. In addition, through its stool-bulking result, fiber could lower the concentration of dirty matters, thereby reducing the number of carcinogen that comes in contact with the gut wall.

6.5.2 Atherosclerosis

The low age-specific incidence of coronary artery disease (CAD) in Central African Republic and parts of India has been attributed partially to a diet characterized by a higher intake of complicated carbohydrates and dietary fibers [75,76]. Plasma steroid alcohol levels are low in these populations; however, different dietary and environmental variations are present. The possible relationship of fiber intake to CHD incidence has additionally been investigated in some populations. In a global survey of nutrient intake and lipids in 1955, it had been ascertained that mean plasma cholesterol for men in Nepal was below that for an age-matched controls [77]. It has been suggested that the massive quantity of fiber from fruit, vegetable, and legumes in Mediterranean-type diets may be partly liable for the low levels of cholesterol in these patients [78]. Low concentrations of lipids and delayed onset of CHD have additionally been reported for vegetarian groups together with Seventh Day Adventists [79], monastic monks [80], strict vegetarians [81], and persons following a Zen vegetarianism [82]. An inverse relationship between cereal fiber intake and death from coronary disease was reported in a retrospective study by Morris et al. [83].

6.5.3 Diabetes

It is commonly assumed that the effect of diets rich in fiber (25 g per diet in women and 38 g per diet in men) [84] can be because of viscous and/or gel-forming characteristics of soluble dietary fibers [85]. Therefore, stronger associations for soluble dietary fibers are expected when the results are analyzed for diets high in active soluble fiber and inactive insoluble fiber separately from prospective cohort studies. However, this hypothesis is not upheld by the present data. A recent metaanalysis of subjects displayed no relation with reduction in the risk of diabetes for fruits and vegetables fiber intake [86,87], but in contrast rich, intake of cereal fibers were markedly related to reduction of the disease.

Evidence from metabolic research supports the impact of a carbohydrate-rich diet compared to fat-rich diet on glycemic response [88–91]. The outcome of ecologic and cross-sectional research support the diminished prevalence of diabetes with heavy intakes of carbohydrate [92–94]; on the contrary, the results of cohort studies do not favor an association between diabetes and total dietary carbohydrates [95–98]. Dietary fiber is documented to reduce postprandial glycemic response and insulin concentrations, by retarding the rate of digestion and absorption of food and by regulating a number of metabolic hormones [99–101].

REFERENCES

[1] Southgate DA. Determination of food carbohydrates. 2nd ed. Norwich (UK): Elsevier; 1991.

[2] Marcason W. What is the latest research regarding the avoidance of nuts, seeds, corn, and popcorn in diverticular disease? J Am Diet Assoc 2008;108(11):1956.

[3] Jenkins AL, Jenkins DJ, Wolever TM, Rogovik AL, Jovanovski E, Božikov V, et al. Comparable postprandial glucose reductions with viscous fiber blend enriched biscuits in healthy subjects and patients with diabetes mellitus: acute randomized controlled clinical trial. Croat Med J 2008;49(6):772–82.

[4] Institute of Medicine (US). Panel on Micronutrients. Dietary Reference Intakes for Energy, Carbohydrate, Fiber, Fat, Fatty Acids, Cholesterol, Protein and Amino Acids. Panel on Macronutrients Panel on the Definition of Dietary Fiber, Subcommittee on Upper Reference Levels of Nutrients, Subcommittee on Interpretation and Uses of Dietary Reference Intakes, and the Standing Committee on the Scientific Evaluation of Dietary Reference Intakes, Food and Nutrition Board. Washington, DC: National Academies Press; 2005.

[5] Southgate DA, Bailey B, Collinson E, Walker AF. A guide to calculating intakes of dietary fibre. J Hum Nutr 1976;30(5):303–13.

[6] Theander O, Aman P. The chemistry, morphology and analysis of dietary fiber components. In: Dietary fibers: chemistry and nutrition. New York, NY: Academic Press; 1979. p. 214–44.

[7] Kay RM. Dietary fiber. J Lipid Res 1982;23(2):221–42.

[8] Southgate DAT, Branch WJ, Hill MJ, Drasar BS, Walters RL, Davies PS, et al. Metabolic responses to dietary supplements of bran. Metabolism 1976;25(10):1129–35.

[9] Pilnik W, Voragen AGJ. Pectic substances and other uronides. In: Hulme AC, editor. Biochemistry of fruits and their products, vol. 1. New York: Academic Press; 1970.

[10] Worth HGJ. The chemistry and biochemistry of pectic substances. Chem Rev 1967;67(4):465–73.

[11] Knuss H. Biosynthesis of pectin and hemicellulose. In: Pridham JB, editor. Plant carbohydrate chemistry. New York, NY: Academic Press; 1974. p. 191–205.

[12] Werch SC, Ivy AC. A study of the metabolism of ingested pectin. Am J Dis Child 1941;62(3):499–511.
[13] Braums FE. The chemistry of lignin. New York, NY: Academic Press; 1952. p. 14–21.
[14] Schubert WJ. Lignin biochemistry. New York: London Academic Press; 1965.
[15] Cherbut C, Salvador V, Barry JL, Doulay F, Delort-Laval J. Dietary fibre effects on intestinal transit in man: involvement of their physicochemical and fermentative properties. Food Hydrocolloids 1991;5(1–2):15–22.
[16] Berg RD. Indigenous intestinal microflora and the host immune response. EOS J Immunol Immunopharmacol 1985;5(4):161–8.
[17] Simone CD, Vesely R, Negri R, Salvadori BB, Zanzoglu S, Cilli A, et al. Enhancement of immune response of murine Peyer's patches by a diet supplemented with yogurt. Immunopharmacol Immunotoxicol 1987;9(1):87–100.
[18] Link-Amster H, Rochat F, Saudan KY, Mignot O, Aeschlimann JM. Modulation of a specific humoral immune response and changes in intestinal flora mediated through fermented milk intake. FEMS Immunol Med Microbiol 1994;10(1):55–63.
[19] Schiffrin EJ, Rochat F, Link-Amster H, Aeschlimann JM, Donnet-Hughes A. Immunomodulation of human blood cells following the ingestion of lactic acid bacteria. J Dairy Sci 1995;78(3):491–7.
[20] Marin ML, Lee JH, Murtha J, Ustunol Z, Pestka JJ. Differential cytokine production in clonal macrophage and T-cell lines cultured with bifidobacteria. J Dairy Sci 1997;80(11):2713–20.
[21] Yasui H, Ohwaki M. Enhancement of immune response in Peyer's patch cells cultured with Bifidobacterium breve. J Dairy Sci 1991;74(4):1187–95.
[22] Holloway WD, Tasman-Jones C, Lee SP. Digestion of certain fractions of dietary fiber in humans. Am J Clin Nutr 1978;31(6):927–30.
[23] Southgate DAT, Durnin JVGA. Calorie conversion factors. An experimental reassessment of the factors used in the calculation of the energy value of human diets. Br J Nutr 1970;24(02):517–35.
[24] Cummings JH, Southgate DAT, Branch WJ, Wiggins HS, Houston H, Jenkins DJA, et al. The digestion of pectin in the human gut and its effect on calcium absorption and large bowel function. Br J Nutr 1979;41(03):477–85.
[25] Kirwan WO, Smith AN, McConnell AA, Mitchell WD, Eastwood MA. Action of different bran preparations on colonic function. Br Med J 1974;4(5938):187–9.
[26] Brodribb AJ, Groves C. Effect of bran particle size on stool weight. Gut 1978;19(1):60–3.
[27] Robertson JA, Eastwood MA. An examination of factors which may affect the water holding capacity of dietary fiber. BrJ Nutr 1981;45(01):83–8.
[28] Story JA, Kritchevsky D. Comparison of the binding of various bile acids and bile salts in vitro by several types of fiber. J Nutr 1976;106(9):1292–4.
[29] Eastwood M, Mowbray L. The binding of the components of mixed micelle to dietary fiber. Am J Clin Nutr 1976;29(12):1461–7.
[30] Kay RM, Strasberg SM, Petrunka CN, Wayman M. Differential adsorption of bile acids by lignins. In: GInglett, Falkehag I, editors. Dietary fibers: chemistry and nutrition. New York, NY: Academic Press; 1979. p. 57–66.
[31] Eastwood MA, Hamilton D. Studies on the adsorption of bile salts to non-absorbed components of diet. Biochim Biophys Acta 1968;152(1):165–73.
[32] Oakenfull DG, Fenwick DE. Adsorption of bile salts from aqueous solution by plant fiber and cholestyramine. Br J Nutr 1978;40(2):299–309.
[33] Malinow MR, Connor WE, McLaughlin P, Stafford C, Lin DS, Livingston AL, et al. Cholesterol and bile acid balance in Macaca fascicularis. Effects of alfalfa saponins. J Clin Invest 1981;67(1):156.
[34] Story JA. Dietary fiber and lipid metabolism: an update. Medical aspects of dietary fiber. New York, NY: Springer; 1980. p. 137–52.

[35] Grant GT, Morris ER, Rees DA, Smith PJ, Thom D. Biological interactions between polysaccharides and divalent cations: the egg-box model. FEBS Lett 1973;32(1):195–8.
[36] McConnell AA, Eastwood MA, Mitchell WD. Physical characteristics of vegetable foodstuffs that could influence bowel function. J Sci Food Agric 1974;25(12):1457–64.
[37] James WPT, Branch WJ, Southgate DAT. Calcium binding by dietary fiber. Lancet 1978;311(8065):638–9.
[38] Sharma BR, Rana V, Naresh L. An overview on dietary fibre. Indian Food Industry 2006;25(5):39–46.
[39] Anderson JW, Bridges SR. Dietary fiber content of selected foods. Am J Clin Nutr 1988;47(3):440–7.
[40] Nordic Nutrition Recommendations 2004. Integrating nutrition and physical activity. 4th ed. Arhus, Denmark: Nordic Council of Ministers; 2005.
[41] GR (Gezondheidsraad). Dietary Reference Intakes: energy, proteins, fats and digestible carbohydrates (Publication no. 2001/19R). The Hague: Health Council of the Netherlands; 2001.
[42] GR (Gezondheidsraad). Guideline for dietary fiber intake (Publication no. 2006/03). The Hague: Health Council of the Netherlands; 2006.
[43] D-A-CH. Referenzwerte für die Nährstoffzufuhr. Deutsche Gesellschaft für Ernährung, Östereichsche Gesellschaft für Ernährung, Schweizerische Gesellschaft für Ernährung, Schweizerische Vereinigung für Ernährung, Umschau Braus, Frankfurt am Main; 2008.
[44] WHO/FAO (World Health Organization/Food and Agriculture Organization). Expert report: diet, nutrition and prevention of chronic diseases: report of a joint WHO/FAO expert consultation (WHO Technical Report Series 916). Geneva: WHO; 2003.
[45] Becker W. Nordic Nutrition Recommendations 2004, based on scientific evidence. Food Nutr Res 2005;49(2):68–71.
[46] DoH (Department of Health). Dietary reference values for food energy and nutrients for the United Kingdom. Report of the Panel on Dietary Reference Values of the Committee on Medical Aspects of Food Policy. London: HMSO; 1991.
[47] Aggett PJ, Agostoni C, Axelsson I, Edwards CA, Goulet O, Hernell O, et al. Nondigestible carbohydrates in the diets of infants and young children: a commentary by the ESPGHAN Committee on Nutrition. J Pediatr Gastroenterol Nutr 2003;36(3):329–37.
[48] Austin MA, Hokanson JE, Edwards KL. Hypertriglyceridemia as a cardiovascular risk factor. Am J Cardiol 1998;81(4):7B–12B.
[49] Trumbo P, Schlicker S, Yates AA, Poos M. Dietary reference intakes for energy, carbohydrate, fiber, fat, fatty acids, cholesterol, protein and amino acids. J Am Diet Assoc 2002;102(11):1621–30.
[50] Pereira MA, Ludwig DS. Dietary fiber and body-weight regulation: observations and mechanisms. Pediatr. Clin. North Am 2001;48(4):969–80.
[51] Cummings JH, Bingham SA, Heaton KW, Eastwood MA. Fecal weight, colon cancer risk, and dietary intake of nonstarch polysaccharides (dietary fiber). Gastroenterology 1992;103:1783–9.
[52] Birkett AM, Jones GP, De Silva AM, Young GP, Muir JG. Dietary intake and faecal excretion of carbohydrate by Australians: importance of achieving stool weights greater than 150 g to improve faecal markers relevant to colon cancer risk. Eur J Clin Nutr 1997;51(9):625–32.
[53] Weaver LT. Bowel habit from birth to old age. J Pediatr Gastroenterol Nutr 1988;7(5):637–40.
[54] Haack VS, Chesters JG, Vollendorf NW, Story JA, Marlett JA. Increasing amounts of dietary fiber provided by foods normalizes physiologic response of the large bowel without altering calcium balance or fecal steroid excretion. Am J Clin Nutr 1998;68(3):615–22.

[55] Stasse-Wolthuis M, Katan MB, Hautvast JG. Fecal weight, transit time, and recommendations for dietary fiber intake. Am J Clin Nutr 1978;31(6):909–10.
[56] Wald A, Scarpignato C, Mueller-Lissner S, Kamm MA, Hinkel U, Helfrich I, et al. A multinational survey of prevalence and patterns of laxative use among adults with self-defined constipation. Aliment Pharmacol Ther 2008;28(7):917–30.
[57] Loening-Baucke V. Chronic constipation in children. Gastroenterology 1993;105(5):1557–64.
[58] Edwards CA, Parrett AM. Dietary fiber in infancy and childhood. Proc Nutr Soc 2003;62(01):17–23.
[59] Nugent AP. Health properties of resistant starch. Nutr Bull 2005;30(1):27–54.
[60] Sharma A, Yadav BS, Ritika. Resistant starch: physiological roles and food applications. Food Rev Int 2008;24(2):193–234.
[61] Mikusova L, Sturdik E, Mosovska S, Brindzova L, Mikulajova A. Development of new bakery products with high dietary fiber content and antioxidant activity for obesity prevention. In: Proceedings of 4th international dietary fiber conference. Vienna, Austria: International association for cereal science and technology (ICC); 2009. 185
[62] Tapsell LC. Diet and metabolic syndrome: where does resistant starch fit in? J AOAC Int 2004;87(3):756–60.
[63] Ranganathan S, Champ M, Pechard C, Blanchard P, Nguyen M, Colonna P, et al. Comparative study of the acute effects of resistant starch and dietary fibers on metabolic indexes in men. Am J Clin Nutr 1994;59(4):879–83.
[64] Tagliabue A, Raben A, Heijnen ML, Deurenberg P, Pasquali E, Astrup A. The effect of raw potato starch on energy expenditure and substrate oxidation. Am J Clin Nutr 1995;61(5):1070–5.
[65] Howe JC, Rumpler WV, Behall KM. Dietary starch composition and level of energy intake alter nutrient oxidation in "carbohydrate-sensitive" men. J Nutr 1996;126(9): 2120.
[66] Raben A, Andersen K, Karberg MA, Holst JJ, Astrup A. Acetylation of or beta-cyclodextrin addition to potato beneficial effect on glucose metabolism and appetite sensations. Am J Clin Nutr 1997;66(2):304–14.
[67] De Roos N, Heijnen ML, De Graaf C, Woestenenk G, Hobbel E. Resistant starch has little effect on appetite, food intake and insulin secretion of healthy young men. Eur J Clin Nutr 1995;49(7):532–41.
[68] Anderson GH, Catherine NL, Woodend DM, Wolever TM. Inverse association between the effect of carbohydrates on blood glucose and subsequent short-term food intake in young men. Am J Clin Nutr 2002;76(5):1023–30.
[69] Keenan MJ, Zhou J, McCutcheon KL, Raggio AM, Bateman HG, Todd E, et al. Effects of resistant starch, a non-digestible fermentable fiber, on reducing body fat. Obesity 2006;14(9):1523–34.
[70] Lopez HW, Levrat-Verny MA, Coudray C, Besson C, Krespine V, Messager A, et al. Class 2 resistant starches lower plasma and liver lipids and improve mineral retention in rats. J Nutr 2001;131(4):1283–9.
[71] Younes H, Levrat MA, Demigné C, Rémésy C. Resistant starch is more effective than cholestyramine as a lipid-lowering agent in the rat. Lipids 1995;30(9):847–53.
[72] Trinidad TP, Wolever TM, Thompson LU. Effect of acetate and propionate on calcium absorption from the rectum and distal colon of humans. Am J Clin Nutr 1996;63(4): 574–8.
[73] Dalgetty DD, Baik BK. Isolation and characterization of cotyledon fibers from peas, lentils, and chickpeas. Cereal Chem 2003;80(3):310–5.
[74] Walker AR. The epidemiological emergence of ischemic arterial diseases. Am Heart J 1975;89(2):133–6.
[75] Rowell H, Burkitt DR. Dietary fiber and cardiovascular disease. Artery 1977;3: 107–19.

[76] Malhotra SL. Serum lipids, dietary factors and ischemic heart disease. Am J Clin Nutr 1967;20(5):462–74.
[77] Keys A, Fidanza F, Keys MH. Further studies on serum cholesterol of clinically healthy men in Italy. Voeding 1955;16:492.
[78] Anderson JT, Grande F, Keys A. Cholesterol-lowering diets. Experimental trials and literature review. J Am Diet Assoc 1973;62(2):133.
[79] Phillips RL, Lemon FR, Beeson WL, Kuzma JW. Coronary heart disease mortality among Seventh-Day Adventists with differing dietary habits: a preliminary report. Am J Clin Nutr 1978;31(10):S191–8.
[80] Barrow JG, Rodiloss PT, Edmands RE, QUINLAN C. Prevalence of atherosclerotic complications in Trappist and Benedictine monks. Circulation 1961;24(4):881.
[81] Hardinge MG, Chambers AC, Crooks H, Stare FJ. Nutritional studies of vegetarians III. Dietary levels of fiber. Am J Clin Nutr 1958;6(5):523–5.
[82] Sacks FM, Castelli WP, Donner A, Kass EH. Plasma lipids and lipoproteins in vegetarians and controls. N Engl J Med 1975;292(22):1148–51.
[83] Morris JN, Marr JW, Clayton DG. Diet and heart: a postscript. Br Med J 1977;2(6098):1307–14.
[84] Howarth NC, Saltzman E, Roberts SB. Dietary fiber and weight regulation. Nutr Rev 2001;59(5):129–39.
[85] Jenkins DJ, Kendall CW, Axelsen M, Augustin LS, Vuksan V. Viscous and nonviscous fibres, nonabsorbable and low glycaemic index carbohydrates, blood lipids and coronary heart disease. Curr Opin Lipidol 2000;11(1):49–56.
[86] de Munter JS, Hu FB, Spiegelman D, Franz M, van Dam RM. Whole grain, bran, and germ intake and risk of type 2 diabetes: a prospective cohort study and systematic review. PLoS Med 2007;4(8). e.261.
[87] Weickert MO, Pfeiffer AF. Metabolic effects of dietary fiber consumption and prevention of diabetes. J Nutr 2008;138(3):439–42.
[88] Swinburn BA, Boyce VL, Bergman RN, Howard BV, Bogardus C. Deterioration in carbohydrate metabolism and lipoprotein changes induced by modern, high fat diet in Pima Indians and Caucasians. J Clin Endocrinol Metab 1991;73(1):156–65.
[89] Garg A, Bantle JP, Henry RR, Coulston AM, Griver KA, Raatz SK, et al. Effects of varying carbohydrate content of diet in patients with non—insulin-dependent diabetes mellitus. JAMA 1994;271(18):1421–8.
[90] Parillo M, Rivellese AA, Ciardullo AV, Capaldo B, Giacco A, Genovese S, et al. A high-monounsaturated-fat/low-carbohydrate diet improves peripheral insulin sensitivity in non-insulin-dependent diabetic patients. Metabolism 1992;41(12):1373–8.
[91] Borkman M, Campbell LV, Chisholm DJ, Storlien LH. Comparison of the effects on insulin sensitivity of high carbohydrate and high fat diets in normal subjects. J Clin Endocrinol Metab 1991;72(2):432–7.
[92] Kawate R, Yamakido M, Nishimoto Y, Bennett PH, Hamman RF, Knowler WC. Diabetes mellitus and its vascular complications in Japanese migrants on the Island of Hawaii. Diabetes Care 1979;2(2):161–70.
[93] Tsunehara CH, Leonetti DL, Fujimoto WY. Diet of second-generation Japanese-American men with and without non-insulin-dependent diabetes. Am J Clin Nutr 1990;52(4):731–8.
[94] Marshall JA, Hamman RF, Baxter J. High-fat, low-carbohydrate diet and the etiology of non-insulin-dependent diabetes mellitus: the San Luis Valley Diabetes Study. Am J Epidemiol 1991;134(6):590–603.
[95] Colditz GA, Manson JE, Stampfer MJ, Rosner B, Willett WC, Speizer FE. Diet and risk of clinical diabetes in women. Am J Clin Nutr 1992;55(5):1018–23.
[96] Salmeron J, Manson JE, Stampfer MJ, Colditz GA, Wing AL, Willett WC. Dietary fiber, glycemic load, and risk of non—insulin-dependent diabetes mellitus in women. JAMA 1997;277(6):472–7.

[97] Salmerón J, Ascherio A, Rimm EB, Colditz GA, Spiegelman D, Jenkins DJ, et al. Dietary fiber, glycemic load, and risk of NIDDM in men. Diabetes Care 1997;20(4):545–50.
[98] Lundgren H, Bengtsson C, Blohme G, Isaksson B, Lapidus L, Lenner RA, et al. Dietary habits and incidence of noninsulin-dependent diabetes mellitus in a population study of women in Gothenburg, Sweden. Am J Clin Nutr 1989;49(4):708–12.
[99] Vinik AI, Jenkins DJ. Dietary fiber in management of diabetes. Diabetes Care 1988;11(2):160–73.
[100] Anderson JW, Akanji AO. Dietary fiber—an overview. Diabetes Care 1991;14(12):1126–31.
[101] Meyer KA, Kushi LH, Jacobs DR, Slavin J, Sellers TA, Folsom AR. Carbohydrates, dietary fiber, and incident type 2 diabetes in older women. Am J Clin Nutr 2000;71(4):921–30.

CHAPTER 2

The Mechanism of Fiber Effects on Insulin Resistance

David Feder, Fernando L.A. Fonseca

1 THE INFLUENCE OF DIET ON METABOLISM

Nutrition plays an important role regarding the development of age-related diseases. Due to the fact that these diseases are associated with higher morbidity and mortality rates, they cause a major impact on public health. A well-balanced diet plan and regular exercise are essential for healthier aging. Throughout the aging process, eating habits may affect the functioning of many organs and tissues, which favorably or unfavorably influence the changes in the body system structures [1].

Protein is a component of utmost importance in a healthy diet. Actually, a great number of studies that focus on the stimulation of protein intake in different phases of life can be found in the literature.

Digestibility is a decisive factor concerning the nutritional quality of the protein in food sources, namely, how easy it is for the body to digest the protein and properly absorb it. Bioavailability, on the other hand, is related to the amount of amino acids the body can effectively make use of after digestion and protein absorption [2].

In general, individuals who base their diets on proteins of vegetal origin and exercise regularly have a lower body mass index. This is due to the lower intake of cholesterol and saturated fat, with a higher proportion of polyunsaturated fatty acids, which directly affects the lipid and glucose profiles of these individuals. Their serum levels of total cholesterol and LDL are systematically lower than in those people with a different kind of diet. Such habits lead to a reduced mortality rate [3].

Proteins of animal origin, milk, and whey proteins highlighted, have high biological levels and bioactive peptides which act not only as antihypertensive, antimicrobial, antioxidant, anticarcinogenic, immune-regulating agents, but also as tumor growth suppressors among other functions [4]. Moreover, dietary fibers also play an important role in the metabolic regulation. Such fibers are the vegetable polysaccharides of the diet, such as cellulose,

Dietary Fiber for the Prevention of Cardiovascular Disease
http://dx.doi.org/10.1016/B978-0-12-805130-6.00002-1

hemicelluloses, pectins, gums, mucilages, and the nonpolysaccharide lignin, which are not hydrolyzed by the human gastrointestinal tract. Some of their important characteristics are as follows: they are of vegetable origin; they are carbohydrates or carbohydrate derivatives (except for lignin); they resist to hydrolysis by digestive enzymes; they are fermentable by colonic bacteria; they reach the colon intact/hydrolyzed, and fermented by the colonic flora.

The combination of animal and vegetable proteins, with the soluble fraction of fibers are beneficial to health given the metabolic effects in the gastrointestinal tract, the gastric emptying and intestinal transit time delay, and the decrease in the absorption of glucose and cholesterol. In a nutritional perspective, the soluble and insoluble fiber intake is recommended in equal parts, totaling 20–30 g on a daily basis (maximum 35 g), or 10–13 g of fibers for each 1000 kcal of energy consumed. Excessive fiber intake affects zinc and calcium absorption, especially in children and elderly.

2 DEFINITION AND IMPORTANCE OF INSULIN RESISTANCE

Insulin resistance is a pathological condition characterized by the impairment of insulin signaling for blood glucose regulation [5]. It is the most important component in a set of factors that act together to trigger the metabolic syndrome [6]. There are epidemiological evidences of the association of insulin resistance not only with glucose intolerance, type 2 diabetes (T2D), hypertension, dyslipidemia, and atherosclerosis but also with many kinds of cancers [7]. It is important to point out that the clinical relevance of insulin resistance cannot be overlooked due to the mortality and morbidity rates related to the many disorders resultant from this condition [8]. It is estimated that 422 million people worldwide suffer from T2D [9]; in the United States alone approximately 29 million people are affected by this condition, and 90% of the cases have T2D among which 90% are obese patients [10]. The prevalence of diabetes more than doubled from 1990 to 2011 in adults of all age groups [11]. Prospective studies have revealed that between 10 and 20 years of age the presence of insulin resistance is detected before the onset of T2D. Therefore, this presence would be the best clinical predictor of a subsequent development of T2D [12,13].

3 PATHOPHYSIOLOGY OF INSULIN RESISTANCE

Insulin resistance is triggered by the complex interaction of genes, obesity, and "the environment" (the latter including nutritional and hormonal factors and the aging process). The rising prevalence of insulin resistance is probably due to the progressive evolution to sedentarism [14]. Interestingly, the genes that have been selected to favor energy effectiveness and storage throughout thousands of years now seem to be highly harmful owing to the excess of nutrients [15]. There are complex mechanisms through which the excess of nutrients leads to insulin resistance, including hormonal imbalance and proinflammatory effects of fat mass increase [16].

In the postfeeding state, amino acids and glucose are driven into the liver by insulin where it stimulates glycogen, protein, and fatty acid synthesis; and suppresses glycogenolysis, gluconeogenesis, proteolysis, and liposis. Insulin also transports glucose and amino acids in the skeletal muscle and glucose into the adipose tissue. It mediates the synthesis of glycogen and glycolysis in the muscle and the synthesis of fatty acid in the adipose tissue. Moreover, insulin also stimulates the synthesis of the lipoprotein lipase in capillaries, a fact that facilitates the extraction of fatty acids from circulation [17,18].

During the fasting hours, insulin levels drop as glucagon levels rise. The decrease in insulin levels promotes glycogenolysis followed by gluconeogenesis in the liver. In the adipose tissue, low levels of insulin stimulate lipolysis, releasing fatty acids to use as fuel. Ketones are generated in the process of fatty acid oxidation. The absence of insulin in the skeletal muscle stimulates proteolysis [17,18].

In the insulin-resistant state, insulin levels are high, but the receptors, especially those in the skeletal muscle, are relatively insensitive to insulin action. High levels of insulin promote the synthesis of fatty acids in the liver. The accumulation and circulation of free fatty acids (FFA) and triglycerides packaged in VLDL aggravate insulin resistance. Thus, the metabolic derangements are self-perpetuating. When β-cells fail, diabetes mellitus develops [17,18].

The main objectives of nutritional management of diabetes are the maintenance of normal or close to normal levels of glucose in serum and the prevention or reversal of lipid abnormalities so that the potential complications of diabetes can be controlled. Nutritional management of insulin resistance aims to prevent the progression of prediabetes to diabetes [19].

4 FIBER—DEFINITION

The term dietary fiber describes the nondigestible carbohydrates and lignin that are intrinsic and intact in plants; functional fiber is the nondigestible carbohydrates with beneficial effects for humans. Total fiber is the sum of dietary and functional fiber [20]. Fiber can still be subdivided into two groups, namely, soluble and insoluble. This grouping is based on chemistry and physics principles, as well as on functional properties [21]. Soluble fiber dissolves in water, forming viscous gels that are not digested in the small intestine. However, when they reach the large intestine, they are easily fermented by its microflora. These gels consist of pectins, gums, fructans like inulin, and some hemicelluloses. Insoluble fiber does not form gels due to its insolubility in water and limited fermentation by intestinal bacteria. Some examples of insoluble fibers are lignin, cellulose, and hemicelluloses [22]. Total dietary fiber intake should be 14 g for 1000 kcal, or 25 g for adult women and 38 g for adult men. Most Americans consume less than optimal intake of fibers, around 16.3 g/day [23].

5 FIBER AND TYPE 2 DIABETES

According to recent studies, higher levels of dietary fiber may contribute to the improvement of the metabolism of carbohydrates in a nonpharmacological way [19].

The association between dietary fiber and T2DM was confirmed in many metaanalyses [19,24–26]. Yao et al. [24] conducted an observational study metaanalysis from 1974 to 2013. Including a total of 488,293 participants, the study showed that in 19,033 cases the risk of T2D decreased with the intake of total dietary, cereal, fruit, and insoluble fiber. Ye et al. [25] analyzed the data of 3,202,850 people/year and revealed a drop in risk of T2D with the consumption of dietary fiber, cereal fiber, and whole grains. Three major cohort studies in the United States showed that individuals with a low cereal-fiber diet and with a high glycemic load are at high risk of developing T2D [26]. In a metaanalysis of 15 studies on patients with T2D, the authors observed lower fasting glycemia and glycated hemoglobin levels when the diet was rich in fiber. The mean fiber intake in the studies was approximately 18 g/day, ranging from 4 to 40 g/day [19]. In Japan, a T2D study that included 4399 patients was carried out through a self-administered questionnaire. The study showed that there was an improvement in the glycemic control associated with a higher fiber intake. Besides other factors, the questionnaire included questions on the history of regular fiber ingestion adjusted according to gender, age, diabetes duration, smoking, tap water consumption, energy expenditure, fat intake, saturated fat intake, physical activities, and the use of oral hypoglycemic drugs or insulin [27].

A different cohort study with older British men suggests that a high-fiber diet (>20 g/day) can potentially reduce the risk of T2D due to the reduction in the inflammatory status [28]. Meyer et al. reported that there was a strong inverse relation of the intake of dietary fiber to diabetes when adjusted to age and BMI [29]. Women who consumed an average of 26 g/day of dietary fiber had a 22% lower risk of developing diabetes when compared with women who only consumed 13 g/day [29]. Schulze et al. corroborated these results in a study with men and women that showed a drop in risk for diabetes ($P < 0.001$) with an additional intake of 12 g of dietary fiber per day. According to these results, focus on the intake of dietary fiber to prevent diabetes may be more significant than on the glycemic index [30].

It is important to observe that the inverse relation of dietary fiber to diabetes shown by Meyer et al. [29] and Schulze et al. [30] did not take age and body weight into consideration. Hu et al. [31] support these conclusions, but adjust the variable age in regard to factors, such as fat intake, smoking, alcohol intake, family history, exercises, and body weight. Therefore, it seems that dietary fibers are related to T2D regardless of other composition factors.

A survey was conducted in South Korea with 3871 healthy adults between 2007 and 2008. It revealed that a diet including whole grains and beans was associated with a significant lower prevalence of insulin resistance [8].

A multicenter, double-blind, randomized study in patients with hypercholesterolemia showed that a diet supplemented with the soluble fiber from *Plantago ovata* husks reduced insulin resistance [32].

An improvement in insulin resistance could also be noted in patients with and without metabolic syndrome treated with dietary fiber (5.59 and 9.49 g of soluble and insoluble fiber, respectively) and L-carnitine [33].

A total of 61 patients with metabolic syndrome were treated with whole grains and compared with a refined grain group. After 12 weeks, the whole grain group had lower postprandial insulin levels [34].

A metaanalysis suggested that the consumption of whole grains, and not refined ones, reduces the risk of T2D [35].

A 5 year evaluation of 175 overweight patients between 8 and 18 years of age revealed that the intake of some specific types of vegetables, even in small quantities, reduces the risk of T2D [36].

6 SOLUBLE AND INSOLUBLE FIBER AND TYPE 2 DIABETES MELLITUS

According to recent studies, the soluble in relation to the insoluble fraction may shed some light on the effectiveness of dietary fiber in diabetes and its mechanisms. The first researches on soluble fiber showed a gastric emptying delay and a reduced absorption of macronutrients, resulting in lower blood glucose and postprandial insulin levels [37]. This most probably occurs due to the viscosity of the soluble fiber in the gastrointestinal tract. Curiously, different types of soluble fibers varied in terms of viscosity and absorption of nutrients. Guar gum was the most viscous and it had the best effect on the reduction of glucose in postprandial blood. Hence, it may be presumed that a higher level of soluble fibers is associated with a lower risk of diabetes. A contradictory hypothesis, nevertheless, since many recent studies have shown no correlation of this sort [29,30,38]. Despite the inconsistencies among studies, showing no differentiation between soluble and insoluble fiber on diabetes [30], most of the researches demonstrate a strong inverse relation between insoluble fiber and T2D. Meyer et al. [29], in a study with healthy middle-aged women, revealed a strong ($P = 0.0012$) inverse relation between insoluble fiber and the risk of T2D whereas soluble fiber showed no effect. Montonen et al. [38] also found the same results in healthy middle-aged men and women who consumed a great amount of whole rye bread. Interestingly, fruit and vegetable fibers had no effect on the risk of developing T2D. Previous studies have agreed with these results. A large epidemiological study including 42,000 men reported that fruit and vegetable dietary fiber had no effect on the risk of diabetes. However, dietary fiber from whole cereal grains showed a significant decrease in diabetes [39]. Daily intakes of fiber in all groups were similar.

7 PATHOPHYSIOLOGY OF THE EFFECT OF FIBER AND TYPE 2 DIABETES MELLITUS

A higher-fiber diet is beneficial in the treatment of T2DM [40], as dietary fiber reduces postprandial hyperglycemia by delaying the digestion and absorption of carbohydrates, and increasing satiety with the effects of a resultant weight loss [19]. In insulin-resistant individuals, dietary fiber can possibly improve peripheral insulin sensitivity through short-chain fatty acids, which are produced by the fermentation of fiber in the intestines [41–43]. The hypoglycemic actions of dietary fiber in patients with T2D were investigated by means of diets rich in fiber or supplementation [19]. For example, a 934-patient epidemiological study in China [44] has recently revealed that HbA1c levels were significantly lower in T2D patients with a fiber-rich diet than in those who followed a western diet [45].

Significant experimental evidences show that the addition of viscous dietary fiber to a diet regimen reduces gastric emptying and digestion rates, as well as glucose absorption. However, the mechanisms through which the fiber affects insulin sensitivity are not clear. An increase in the consumption of dietary fiber, particularly that found in low glycemic index foods, may reduce the risk of T2D [23].

Insoluble fiber plays a small role regarding the absorption of macronutrients [37]. Therefore, a different course of action must be established and many hypotheses should be discussed. Some of these hypotheses suggest that insoluble fiber increases the food passage rate through the GI tract, thus resulting in a decrease in the absorption of nutrients, namely, simple carbohydrates. However, Weicket et al. [46] found that an increase in the intake of cereal fiber significantly improved whole body glucose disposal, resulting in an 8% improvement of insulin sensitivity. This suggests that the underlying mechanisms of the insoluble fiber are more peripheral, and they are not limited to the absorption of nutrients. First, accelerated gastric inhibitory polypeptide (GIP) release was observed immediately after the ingestion of insoluble fiber by healthy women [47]. GIP is an incretin hormone that mediates postprandial insulin release. Second, insoluble fiber may lead to reduced consumption of food and appetite [48], which results in a caloric intake reduction. By means of fermentation, tertiary fatty acid (short chain) reduced postprandial glucose response [49,50]. The first studies showed that lipid administration harms glucose metabolism [51] and that oral acetates may reduce FFA levels in the blood [52]. According to Kelley and Mandarino [53], the increase of FFA levels in the blood may hinder glucose metabolism by the inhibition of GLUT4 glucose transporters. Therefore, short-chain fatty acids decrease FFA levels in plasma, which may reduce glucose levels in the blood through the competition in tissues that are sensitive to insulin.

The inverse relation between cereal grains and diabetes can also be related to an increase in levels of magnesium, which is administered to reduce the incidence of T2D [29,54]. Hypomagnesemia is common in diabetic patients, and it has been associated with reduced tyrosine kinase activity of the insulin receptor [55], which may hinder the action of insulin and predispose the patient to insulin resistance.

Finally, it appears that the action of the fibers with the intention of reversing or even as an alternative to mitigate metabolic disorders is vast. The fiber interference actions can be numerous and because of that more studies are needed to verify the mechanisms by which the fibers can act minimizing non-communicable chronic diseases.

REFERENCES

[1] Vellas B. Nutrition and health in an ageing population. Belfield, Dublin, Ireland: UCD Institute of Food and Health; 2009.
[2] Pires KMP, Mandarim-De-Lacerda CA. Restrição protéica na lactação como causa de hiperplasia da túnica media da aorta em ratos adultos. Rev. SOCERJ 2005;18(3): 193–200.
[3] Couceiro P, Slywitch E, Lenz F. Padrão alimentar da dieta vegetariana. Einstein 2008;6(3):365–73.
[4] Parodi PW. A role for milk proteins and their peptides in cancer prevention. Curr Pharm Des 2007;13(8):813–28.
[5] Djiogue S, Nwabo Kamdje AH, Vecchio L, Kipanyula MJ, Farahna M, Aldebasi Y, et al. Insulin resistance and cancer: the role of insulin and IGFs. Endocr Relat Cancer 2013;20(1):R1–17.
[6] Karagiannis T, Paschos P, Paletas K, Matthews DR, Tsapas A. Dipeptidyl peptidase-4 inhibitors for treatment of type 2 diabetes mellitus in the clinical setting: systematic review and meta-analysis. BM J 2012;12:1344–69.
[7] Song S, Paik HY, Song Y. High intake of whole grains and beans pattern is inversely associated with insulin resistance in healthy Korean adult population. Diabetes Res Clin Pract 2012;98(3):e28–31.
[8] Global Report of Diabetes Mellitus Data generated by World Health Organization and available at http://www.who.int/diabetes/en/
[9] http://www.diabetes.org/diabetes-basics/statistics/
[10] http://www.who.int/mediacentre/factsheets/fs312/en/
[11] Warram JH, Martin BC, Krolewski AS, Soeldner JS, Kahn CR. Slow glucose removal rate and hyperinsulinemia precede the development of type II diabetes in the offspring of diabetic patients. Ann Intern Med 1990;113:909–15.
[12] Lillioja S, Mott DM, Howard BV, Bennett PH, Yki-Järvinen H, Freymond D, et al. Impaired glucose tolerance as a disorder of insulin action. Longitudinal and cross-sectional studies in Pima Indians. N Engl J Med 1988;318:1217–25.
[13] Lebovitz H. Type 2 diabetes: an overview. Clin Chem 1999;45:1339–45.
[14] Spiegelman BM, Flier JS. Obesity and the regulation of energy balance. Cell 2001;104:531–43.
[15] Jung UJ, Choi MS. Obesity and its metabolic complications: the role of adipokines and the relationship between obesity, inflammation, insulin resistance, dyslipidemia and nonalcoholic fatty liver disease. Int J Mol Sci 2014;15(4):6184–223.
[16] Ye J. Role of insulin in the pathogenesis of free fatty acid induced insulin resistance in skeletal muscle. Endocr Metab Immune Disord Drug Targets 2007;7:65–74.
[17] Hawkins M, Rossetti L. Insulin resistance and its role in the pathogenesis of type 2 diabetes. In: Kahn R, Weir G, King G, Jacobson A, Smith R, Moses A, editors. Joslin's Diabetes Mellitus. 14th ed. Boston: Lippincott, Williams and Wilkins; 2005. p. 425–48.
[18] Sesti G. Pathophysiology of insulin resistance. Best Pract Res Clin Endocrinol Metab 2006;20:665–79.
[19] Post RE, Mainous AG, King DE, Simpson KN. Dietary fiber for the treatment of type 2 diabetes mellitus: a meta-analysis. J Am Board Fam Med 2012;25:16–23.
[20] Institute of Medicine, Food Nutrition Board. Dietary reference intakes for energy, carbohydrates, fiber, fat, fatty acids, cholesterol, protein, and amino acids. Washington, DC: The National Academies Press; 2005.
[21] Sizer F, Whitney E. Nutrition: concepts and controversies. 11th ed. Belmont, CA: Thomson Wadsworth; 2008.
[22] Wong JM, Jenkins DJ. Carbohydrate digestibility and metabolic effects. J Nutr 2007;137:2539S–46S.

[23] Dahl WJ, Stewart ML. Position of the academy of nutrition and dietetics: health implications of dietary fiber. J Acad Nutr Diet 2015;115:1861–70.
[24] Yao B, Fang H, Xu W, et al. Dietary fiber intake and risk of type 2 diabetes: a doseresponse analysis of prospective studies. Eur J Epidemiol 2014;29(2):79–88.
[25] Ye EQ, Chacko SA, Chou EL, Kugizaki M, Liu S. Greater whole-grain intake is associated with lower risk of type 2 diabetes, cardiovascular disease, and weight gain. J Nutr 2012;142(7):1304–13.
[26] Bhupathiraju SN, Tobias DK, Malik VS, Pan A, Hruby A, Manson JE, et al. Glycemic index, glycemic load, and risk of type 2 diabetes: results from 3 large US cohorts and an updated meta-analysis. Am J Clin Nutr 2014;100(1):218–32.
[27] Fujii H, Iwase M1, Ohkuma T, Ogata-Kaizu S, Ide H, Kikuchi Y, et al. Impact of dietary fiber intake on glycemic control, cardiovascular risk factors and chronic kidney disease inJapanese patients with type 2 diabetes mellitus: the Fukuoka Diabetes Registry. Nutr J 2013;12:159.
[28] Wannamethee SG, Whincup PH, Thomas MC, Sattar N. Associations between dietary fiber and inflammation, hepatic function, and risk of type 2 diabetes in older men: potential mechanisms for the benefits of fiber on diabetes risk. Diabetes Care 2009;32(10):1823–5.
[29] Meyer KA, Kushi LH, Jacobs DR, Slavin J, Sellers TA, Folsom AR. Carbohydrates, dietary fiber, and incident type 2 diabetes in older women. Am J Clin Nutr 2000;71: 921–30.
[30] Schulze MB, Liu S, Rimm EB, Manson JE, Willett WC, Hu FB. Glycemic index, glycemic load, and dietary fiber intake and incidence of type 2 diabetes in younger and middle-aged women. Am J Clin Nutr 2004;80:348–56.
[31] Hu FB, Manson JE, Stampfer MJ, Colditz G, Liu S, Solomon CG, et al. Diet, lifestyle, and the risk of type 2 diabetes mellitus in women. N Engl J Med 2001;345: 790–7.
[32] Solà R, Bruckert E, Valls RM, Narejos S, Luque X, Castro-Cabezas M, et al. Soluble fibre (*Plantago ovata* husk) reduces plasma low-density lipoprotein (LDL) cholesterol, triglycerides,insulin, oxidised LDL and systolic blood pressure in hypercholesterolaemic patients: a randomised trial. Atherosclerosis 2010;211(2):630–7.
[33] Banuls C, Rovira-Llopis S, Monz N, Sol_E, Vadel B, Víctor VM, et al. The consumption of a bread enriched with dietary fibre and L-carnitine improves glucose homoeostasis and insulin sensitivity in patients with metabolic syndrome. J Cereal Sci 2015;64: 159–67.
[34] Giacco R, Costabile G, Della Pepa G, Anniballi G, Griffo E, Mangione A, et al. A whole-grain cereal based diet lowers postprandial plasma insulin and triglyceride levels in individuals withmetabolic syndrome. Nutr Metab Cardiovasc Dis 2014;24(8): 837–44.
[35] Aune D, Norat T, Romundstad P, Vatten LJ. Whole grain and refined grain consumption and the risk of type 2 diabetes: a systematic review and dose–response meta-analysis of cohort studies. Eur J Epidemiol 2013;28:845–58.
[36] Cook LT, O'Reilly GA, Goran MI, Weigensberg MJ, Spruijt-Metz D, Davis JN. Vegetable consumption is linked to decreased visceral and liver fat and improved insulin resistance in overweight Latino Youth. J Acad Nutr Diet 2014;114:1776–83.
[37] Jenkins DJ, Wolever TM, Leeds AR, Gassull MA, Haisman P, Dilawari J, et al. Dietary fibres, fibre analogues, and glucose tolerance: importance of viscosity. Br Med J 1978;1:1392–4.
[38] Montonen J, Knekt P, Jarvinen R, Aromaa A, Reunanen A. Whole-grain and fiber intake and the incidence of type 2 diabetes. Am J Clin Nutr 2003;77:622–9.
[39] Salmeron J, Ascherio A, Rimm EB, Colditz GA, Spiegelman D, Jenkins DJ, et al. Dietary fiber, glycemic load, and risk of NIDDM in men. Diabetes Care 1997;20:545–50.

[40] American Diabetes Association,. Standards of medical care in diabetes—2013. Diabetes Care 2013;36(Suppl. 1):S11–66.
[41] Johnston KL, Thomas EL, Bell JD, Frost GS, Robertson MD. Resistant starch improves insulin sensitivity in metabolic syndrome. Diabet Med 2010;27:391–7.
[42] Robertson MD, Wright JW, Loizon E, Debard C, Vidal H, Shojaee-Moradie F, et al. Insulin-sensitizing effects on muscle and adipose tissue after dietary fiber intake in men and women with metabolic syndrome. J Clin Endocrinol Metab 2012;97:3326–32.
[43] Weickert MO, Mohlig M, Schofl C, Arafat AM, Otto B, Viehoff H, et al. Cereal fiber improves whole-body insulin sensitivity in overweight and obese women. Diabetes Care 2006;29:775–80.
[44] Jiang J, Qiu H, Zhao G, Zhou Y, Zhang Z, Zhang H, et al. Dietary fiber intake is associated with HbA1c level among prevalent patients with type 2 diabetes in Pudong New Area of Shanghai China. PLoS One 2012;7:e46552.
[45] Zhou BF, Stamler J, Dennis B, Moag-Stahlberg A, Okuda N, Robertson C, et al. Nutrient intakes of middle-aged men and women in China, Japan, United Kingdom, and United States in the late 1990s: the INTERMAP study. J Hum Hypertension 2003;17:623–30.
[46] Weickert MO, Pfeiffer AF. Metabolic effects of dietary fiber consumption and prevention of diabetes. J Nutr 2008;138:439–42.
[47] Weickert MO, Mohlig M, Koebnick C, Holst JJ, Namsolleck P, Ristow M, et al. Impact of cereal fibre on glucose-regulating factors. Diabetologia 2005;48:2343–53.
[48] Samra R, Anderson GH. Insoluble cereal fiber reduces appetite and short-term food intake and glycemic response to food consumed 75 min later by healthy men. Am J Clin Nutr 2007;86:972–9.
[49] Brighenti F, Castellani G, Benini L, Casiraghi MC, Leopardi E, Crovetti R, et al. Effect of neutralized and native vinegar on blood-glucose and acetate responses to a mixed meal in healthy-subjects. Eur J Clin Nutr 1995;49:242–7.
[50] Ostman EM, Liljeberg Elmstahl HG, Bjorck IM. Barley bread containing lactic acid improves glucose tolerance at a subsequent meal in healthy men and women. J Nutr 2002;132:1173–5.
[51] Ferrannini E, Barrett EJ, Bevilacqua S, DeFronzo RA. Effect of fatty acids on glucose production and utilization in man. J Clin Invest 1983;72:1737–47. Nutrients 2010;2:1285.
[52] Crouse JR, Gerson CD, Decarli L, Lieber CS. Role of acetate in the reduction of plasma free fatty acids produced by ethanol in man. J Lipid Res 1968;9:509–12.
[53] Kelley DE, Mandarino LJ. Fuel selection in human skeletal muscle in insulin resistance—a reexamination. Diabetes 2000;49:677–83.
[54] Slavin JL, Martini MC, Jacobs DR Jr, Marquart L. Plausible mechanisms for the protectiveness of whole grains. Am J Clin Nutr 1999;70:459S–63S.
[55] Paolisso G, Barbagallo M. Hypertension, diabetes mellitus, and insulin resistance: the role of intracellular magnesium. Am J Hypertension 1997;10:346–55.

FURTHER READING

Calbet JAL, Holst JJ. Gastric emptying, gastric secretion and enterogastrone response after administration of milk proteins or their peptide hydrolysates in humans. Eur J Nutr 2004;43(3):127–39.
Chagas EFB, et al. Exercício físico e fatores de risco cardiovasculares em mulheres obesas na pós-menopausa. Rev Bras Med Esporte, São Paulo 2015;21(1):65–9.
Ciolac EG, Brech GC, Greve JMD. Age does not affect exercise intensity progression among women. J Strength Cond Res 2010;24(11):3023–31.

da Silva Junior BA, et al. Desempenho de vacas leiteiras alimentadas com cana de açúcar associada à ureia e tratada com cal virgem na região do Alto Pantanal Sul-Mato-Grossense. Semina: Ciências Agrárias 2015;36(3 1):2317–28.

Farr JN, et al. Progressive resistance training improves overall physical activity levels in patients with early osteoarthritis of the knee: a randomized controlled trial. Phys Ther 2010;90(3):356–66.

Fernandez AC, Mello MT, Tufik S, Castro PM, Fisberg M. Influência do Treinamento Aeróbio e Anaeróbio na Massa de Gordura de Adolescentes Obesos. Rev Bras Med Esporte 2004;10(3):152–8.

Fleck SJ, Kraemer WJ. Fundamentos do Treinamento de Força Muscular. Editora Artmed. 2nd ed. Porto Alegre, 1999.

Glaucia FB, et al. Morphological and biochemical effects on the skeletal muscle of ovariectomized old female rats submitted to the intake of diets with vegetable or animal protein and resistance training. Oxid Med Cell Longev 2016;2016:1–10.

Hauser C, Benetti M, Rebelo FPV. Estratégias para o emagrecimento. Rev Bras Cineantropom Desempenho Hum 2004;6(1):72–81.

Marchon C, et al. Effects of moderate exercise on the biochemical, physiological, morphological and functional parameters of the aorta in the presence of estrogen deprivation and dyslipidemia: an experimental model. Cell Physiol Biochem 2015;35(1):397–405.

Mediano MFF, Paravidino V, Simão R, Polito MD, Pontes FL. Comportamento subagudo da pressão arterial após o treinamento de força em hipertensos controlados. Rev Bras Med Esporte 2005;111(6):240–337.

Negrão CE, Barreto ACP. Efeito do treinamento físico na insuficiência cardíaca: implicações autonômicas, hemodinâmicas e metabólicas. Revista Soc Cariol de SP 1998;8(2).

Nieman DC. Exercício e Saúde: como prevenir de doenças usando o exercício como seu medicamento. Ed Manole, São Paulo, 1999.

Sabia RV, Santos E, Ribeiro RPP. Efeito da Atividade Física Associada à Orientação Alimentar em Adolescentes Obesos: comparação entre o exercício aeróbio e anaeróbio. Rev Bras Med Esporte 2004;10(5):349–55.

Schmitz KH, Jensen MD, Kugler KC, Jeffery RW, Leon AS. Strengthtraining for obesity prevention in midlife women. Int J Obes Relat Metab Disord 2003;27:326–33.

Winett RA, Carpinelli RN. Potential health-related benefits of resistance training. Prev Med 2001;33(5):503–13.

CHAPTER 3

The Interaction Between Insoluble and Soluble Fiber

Deepak Mudgil

1 INTRODUCTION

In 1953, Hispley coined the term "Dietary fiber" for first time, which included the cell wall portions, such as cellulose, hemicelluloses, and lignin [1]. In 430 BC, Hippocrates explained the laxative action of coarse wheat against refined wheat. In 1920, J. H. Kellogg reported attributes of bran, such as increased stool weight and enhanced laxation. After this, very little or no research work was carried out on dietary fiber until the 1970s [2]. Generally, dietary fibers are indigestible in human small intestine and are considered as "roughage" materials that are resistant to digestion by the secretions of the human digestive tract. If these nondigestible carbohydrates are isolated from the plant sources and have physiological health benefits, then these are known as functional fibers. Classification of dietary fiber can be done on the basis of their solubility, fermentability, and physiological effects. The term dietary fibers generally include nonstarch polysaccharides (NSP), oligosaccharides, lignin, and associated plant substances (such as cutin, suberin, phytic acid, and so on) [3]. Resistant starches (RS) are also included in the definition of dietary fiber as they also resist digestion and not absorbed in small intestine of healthy individuals [4]. Whole grains include partly milled grains, breakfast cereals, such as All Bran or Fiber One. In the last decade, dietary fibers are extensively studied for their participation in regulation of physiological functions, such as lowering of blood cholesterol, improvements in gastrointestinal mobility(GI), regulation of glucose and lipid metabolism, stimulation of bacterial metabolic activity, detoxification of colon luminal content, and integrity of intestinal mucosa. When dietary fiber components are consumed in the diet, they also interact with the other food components and can potentially interfere with metabolism and absorption. List of dietary fiber components along with their food sources are given in Table 3.1.

Dietary Fiber for the Prevention of Cardiovascular Disease
http://dx.doi.org/10.1016/B978-0-12-805130-6.00003-3

Table 3.1 Dietary fiber components from food sources

Food source	Dietary fiber component
Cereals	Cellulose
	Xylan
	Arabinoxylan
	β-D-Glucan
	Fructans
	Galactans
	Lignin and lignan
	Phenolic ester
	NSP
Legumes	Galactomannans
	Glucomannan
	Xyloglucans
	Pectin
	Lignin
	Cellulose
	NSP
Fruits and vegetables	Cellulose
	Pectin
	Xyloglucans
	Lignins
	Cutin
	Waxes
	Gums and mucilages

NSP, Nonstarch polysaccharides.

2 CLASSIFICATION OF DIETARY FIBERS

Several classification systems have been suggested for the classification of dietary fiber components, which are based on their role, fiber constituents, type of polysaccharide, GI solubility, site of digestion, and products of digestion and physiological classification. However, none of these systems are solely satisfactory because they do not cover all the aspects completely. Among these systems of classification are the two most accepted classifications, which are based on the concept of solubility and on the concept of fermentability of dietary fiber in an in-vitro system using an aqueous enzyme solution representative of human alimentary enzymes. Generally, it has been considered that fermented fibers are soluble in water, whereas nonfermented or poorly fermented fibers are insoluble in water [5].

According to its solubility, there are two different categories of fibers present in dietary constituents: insoluble dietary fibers and soluble dietary

Table 3.2 Different classes of dietary fiber

Dietary fibers		
Insoluble dietary fibers	**Soluble dietary fibers**	
Poorly fermentable or non-fermentable fibers	**Fermentable fibers**	
Celluloses	*Viscous*	*Nonviscous*
Hemicelluloses	Pectin	Resistant dextrin
Lignins	β-D-Glucan	Resistant starch
Resistant starches	Galactomannans	Polydextrose
Arabinoxylans	Glucomannans	Inulin
NSP	Psyllium	FOS

FOS, Fructooligosaccharides; NSP, nonstarch polysaccharides.

fibers. Both types of fibers have different actions and influence on the normal gut activity. Soluble fiber dissolves in water whereas insoluble fiber does not dissolve in water. Soluble dietary fibers bypass the digestion process of the small intestine and are easily fermented by the microflora of the large intestine. Sources of soluble dietary fiber include pectin, gums, starches, oats, psyllium, fruit, vegetables, pulses, etc. [6]. Soluble fibers upon dissolution in water may or may not have viscosity effects, which means that after dissolution in water it forms a viscous gel. This characteristic of soluble dietary fiber further classifies them into two categories—viscous fibers and nonviscous fibers. Viscous soluble fiber includes pectin, galactomannan, glucomannan, β-glucan, psyllium, etc. whereas nonviscous soluble fiber includes fructooligosaccharides (FOS), resistant dextrin, inulin, etc. (Tables 3.2 and 3.3).

Table 3.3 Classification of dietary fiber based on solubility/fermentability

Property	**Component**	**Sources**
Water solubility/fermentability	Pectin	Fruits and vegetables, legumes, etc.
	Gums	Guar seeds, locust bean, seaweeds extract, microbial gums, etc.
	Mucilages	Vegetables, plant extracts, etc.
Water insolubility/little or no fermentability	Cellulose	Cereal brans, such as rice bran, wheat bran, root vegetables, peas, beans, legumes, etc.
	Hemicellulose	Cereal whole grains, cereal bran, pulses, dried beans, etc.
	Lignin	Cereals, stone fruits, edible seeds, vegetable filaments, flaxseed, etc.

Soluble fibers generally increase the transit time through human digestive tract, which results in delayed gastric emptying, which ultimately results in slower glucose absorption. In the human GI tract, insoluble dietary fibers are not soluble in water and do not form gels due to their water insolubility and no or very limited fermentation of insoluble fiber occurs in the large intestine. Cereal brans, such as All Bran, Fiber One, and rice bran and whole grains are major sources of insoluble dietary fiber in the form of cellulose and lignins. Wheat bran generally contains up to 50% dietary fiber while oat bran reported to have about 20% dietary fiber [4]. Insoluble fibers generally increase the fecal bulk and excretion of bile acids, while decreasing intestinal transit time. Insoluble and soluble fibers are present in different concentration in various foods and have different properties; hence it is important to include a variety of fiber-containing foods in daily dietary plan. Besides beneficial physiological effects, such as cholesterol lowering, diabetes control, and digestive system improvements, dietary fibers also improves the growth and activity of beneficial bacteria in human digestive tract by acting as food for the beneficial human intestinal microflora. This activity of dietary fiber is known as prebiotic activity and the fibers showing this activity are known as prebiotics. Prebiotic are generally defined as the indigestible food component that beneficially influences the host organism by selective stimulation of growth and activity of beneficial bacteria (such as Lactobacilli and Bifidobacteria) in the human digestive tract which leads to improvement in the host health. Examples of prebiotic substance includes guar gum, gum acacia, tragacanth gum, FOS, and galactooligosaccharides which act as food for the human colonic bacteria and helps in growth and activity of these bacteria [7]. Insoluble dietary fiber is more abundant in foods as compared to soluble dietary fiber. Most of the fiber containing foods generally contains approximately one-third soluble and two-third insoluble fiber [8].

3 PROPERTIES OF DIETARY FIBER

The physiological effects caused by dietary fibers are dependent on their physicochemical properties, such as solubility, fermentability, viscosity, water absorption and water-holding capacity, adsorption and binding ability, and particle size [9].

3.1 Solubility

Dietary fibers are of two types depending on their solubility in water: soluble and insoluble dietary fiber. The soluble and insoluble nature of dietary fibers influences their physiological effects and technological functions [10]. When ingested in diet, soluble fibers increase the viscosity of the aqueous phase and leads to reduction in the glycemic response and plasma cholesterol [11,12]. Insoluble fibers are designated by their porosity and low density. Due to these properties, insoluble dietary fiber results in increase in fecal bulk and decrease in intestinal transit [13]. In functional foods market, development of soluble fiber fortified food products is more beneficial as it provides viscosity and ability to form gels, as compared to insoluble fiber [6].

3.2 Fermentability

Fermentability is an important characteristic of the dietary fiber as it can contribute as substrate for fermentation. It has been reported that the average excretion of dry weight and energy on low fiber diet are 50 g/day and 800 kJ/day, respectively and 88 g/day and 1700 kJ/day, respectively on a high fiber diet [14]. Fermentation of the fiber is highly variable including little or no fermentation (lignins) to almost complete fermentation (pectins). Among dietary fibers, soluble fibers are more prone to fermentation by colonic bacteria as compared to insoluble fibers [15]. Fermentability of dietary fiber is directly related with its effect on bowel function, such as fecal mass, stool frequency, maintenance of colonic pH, and retrieve energy from nondigestible foods are directly related to their fermentation pattern [16].

Depending on the fermentability, dietary fibers are classified as rapidly fermented, slowly fermented, and nonfermentable dietary fibers. Fruits, such as apples and vegetables, such as beans are considered as rapidly fermented proportions and these may contribute less to fecal bulking than other fibers. Psyllium and wheat bran are considered as slow fermenting proportions and contribute to fecal mass through fermentation. Dietary fibers with poor fermentation capacity contribute toward increased bulk in the large intestine, which results in reduced risk of constipation and colonic cancer. Dietary fibers with high fermentability are associated with physiological actions on colonic mucosa via production of fermentation metabolites [16,17]. Highly fermented fibers are also associated with postabsorptive actions on the liver and other tissues [18].

In addition to these functions, some fiber sources (FOS) can selectively stimulate the growth and activity health-promoting bacteria, such as bifidobacteria and lactobacilli which are responsible for the protective barrier function and for stimulating healthy immune response in adults [19].

The access of bacteria to substrate is linked with the physical structure of the fiber matrix and the chemical structure of dietary fiber component, which ultimately describe the rate and extent of fermentation [9]. Fiber components, which contain high amounts of secondary tissues, such as hull and bran are poorly fermented compared to fiber products rich in parenchyma tissue, such as fruits and vegetables. Soluble polysaccharides exhibits higher fermentation rate than equivalent polysaccharides within cell walls. Blocked branched polysaccharides exhibit more fermentation as compared to highly and randomly branched polysaccharides. The course of fermentation depends on the microflora composition of the individual. It may vary from one individual to another individual and also within the cell population densities of the principal taxonomic groups [20]. The examples of bacteria that can hydrolyze polysaccharides include *Bacteroides*, *Bifidobacterium*, *Ruminococcus*, and some species of *Eubacterium* and *Clostridium*. Some substrates, such as FOS and inulin can promote selectively the growth and/or the activity of one or a limited number of bacteria in the colon and are known as prebiotics. End products of fermentation include short chain fatty acids (SCFA), mainly acetate, propionate, and butyrate and gases.

3.3 Viscosity

Viscosity is one of the significant properties associated with soluble dietary fibers [21]. Some soluble dietary fiber, such as pectins, gums, psyllium, and β-glucan may form viscous solutions when interact with aqueous phase. Viscosity or gel-formation is associated with the ability of dietary fiber to absorb water, which results in the formation of gelatinous mass. Inclusion of these dietary fibers in daily diet can enhance the volume, as well as viscosity of the contents of the GI tract [13]. This increase in volume and viscosity of digesta leads to delay gastric emptying in the stomach, which ultimately can enhance satiety. This enhanced viscosity also reduces the emulsification of dietary lipids in the acid medium of the stomach, which results in lower extent of lipid assimilation. Viscosity caused by the action dietary fiber can resist the effects of GI motility in the lumen of small intestine. Because of the enhanced viscosity, gut content behaves like gel and responds more like solids than liquids in the GI tract. This phenomenon can describe the delayed gastric emptying often associated with the ingestion of fibers due to the reduced diffusion of digestive enzymes toward their substrates, which ultimately slows down the digestion. Enhanced viscosity of the digested contents can also retard the release and transit of the hydrolysis products toward the absorptive surface of the mucosa. Chemical structure of the dietary fiber component and its interaction with other macromolecules are responsible for the viscosity. Volume occupied by the dietary fiber components are generally characterized by its intrinsic viscosity [22].

This viscosity property of the dietary fiber components can be enhanced or reduced by certain treatments. Hydrolysis treatments of dietary fiber components, such as gums, pectins, and beta glucans into lower molecular weight molecules reduce the viscosity enhancement capacity of the digestive contents by these fiber components. Certain treatments, such as extrusion cooking can enhance the amount of water-soluble dietary fiber component and ultimately the viscosity caused by dietary fiber.

Viscosity of the digestive contents containing dietary fiber, do not remain the same throughout the gut. Dietary fiber components are generally not digested by enzymes in upper gut, but they can undergo significant degradation, for example, pectin can be solubilized via breakdown of calcium bridges in stomach at acidic pH or via beta elimination in small intestine at about neutral pH. The extent of this degradation of dietary fiber component could have a nutritional impact due to reduced viscosity action which is further associated with physiological actions. Solubility and viscosity of polysaccharides can also be influenced via any change in ionic environment and pH throughout the GI tract. This phenomenon is generally observed in case of polyelectrolytes, such as alginate, which forms a gel in the stomach but solubilizes in the small intestine. The viscosity of fluid digesta may vary within the gut, which means that the measured viscosity of a dietary fiber source may have little relationship to the viscosity in the digestive segments of interest depending on the conditions in the gut and type of dietary fiber. Concentration, structure, and molecular weight of the polymer under consideration are important to study the effect of particular dietary fiber source. Viscosity value for dietary fiber sources should be reported at a range of values of the shear rates as most of the dietary fibers show shear thinning behavior. Hence, viscosity value at a single shear rate cannot explain the complete rheological behavior of the dietary fiber [23].

3.4 Water Absorption and Water-Holding Capacity

Insoluble type of dietary fibers, such as lignin and cellulose are mostly unfermentable by colonic microflora and increase fecal bulk by swelling and water-holding capacity [15]. Water holding or water retention capacity is described by the amount of water that is retained by known weight of dry fibers under specified conditions of temperature, time of soaking, and duration of centrifugation and centrifugation speed. A portion of the soluble fiber is generally lost during measurement, which affects the water-holding capacity; hence the amount of water measured by centrifugation is generally higher than the amount of water absorbed [24].

Polysaccharide components of dietary fibers are generally hydrophilic in nature and the water molecule is adhered on the hydrophilic sites of the fiber component or within the intercellular void spaces in the molecular structure. Hence, insoluble dietary fibers can absorb, swell, and entrap water within its porous matrix in its molecular structure, which results in the bulking effect of fiber in the colon. Insoluble dietary fiber components can reduce the potency of cytotoxic substances via dilution of such substances in the large intestine. The hydration properties of dietary fiber components depend on the chemical nature of the components, their arrangement in the cell walls and the anatomy, and particle size of the dietary fiber. The dietary fiber components, which are composed of primary cell walls shows higher hydration capacity as compared to the fiber components, composed of secondary cell walls [25,26]. Hydration properties of fiber components containing charged groups, such as pectin are influenced by pH, ionic strength, and nature of the ions. Processing treatments, such as heating, grinding, extrusion, and so on given to food components containing fiber components affects the hydration properties of dietary fiber component via modification of composition and physical properties of fiber matrix.

3.5 Adsorption and Binding Ability

Trapping of bile acids or adsorption of bile acids has been suggested as a potential mechanism by which certain dietary fiber components may increase the excretion of bile acids in feces. Soluble fiber forms gel matrices, which are eventually excreted in the feces entrap some of the bile acids released from the gallbladder. This physical entrapment of bile acids appears to be more pronounced in the terminal region of ileum where reabsorption of bile acids from the digesta usually occurs [15]. Enhanced excretion of bile acids leads to higher cholesterol turnover from the body. The exact mechanisms of binding of bile acids by dietary fiber components are still not known whereas hydrophobic and ionic interactions have been suggested in support of the mechanism [25]. Dietary fiber components composed of lignified or coarse tissues (e.g., rice straw) are reported to have binding properties [27]. Adsorption of the components can be measured by methods which are similar to those generally used for water retention capacity. Retention of the components is caused by their absorption and entrapment in cell matrix. Physiologically reliable measurement of binding capacity of dietary fiber components is based on certain conditions, such as chemical environment of the small intestine and behavior of dietary fiber components in small intestine.

3.6 Particle Size

Particle size of dietary fiber is an important characteristic, which indirectly control the physiological function in the digestive tract, such as transit time, fermentation, and fecal excretion. The rate of fermentation of dietary fibers is directly related to surface area of the fiber, which is in direct contact with bacteria [28]. Wheat bran with coarser particle size is more effective in regulating transit time as compared to wheat bran with fine particle size. Dietary fiber components reduce the intestinal transit time, which is beneficial in terms of protection of colon from the extended exposure to cytotoxic substances, which may be harmful for human health. Particle size of dietary fiber depends on processing treatments on the fiber product. Mechanical treatment, such as grinding, as well as chewing reduces the particle size of the dietary fiber. Approximate complete disintegration of the particles can be achieved by degradation of the fiber matrix by colonic bacteria.

4 SOLUBLE FIBER INTERACTIONS

4.1 β-Glucan

β-Glucan is a water-soluble dietary fiber obtained from oats, barley, bacteria, yeast, algae, and mushrooms [29]. Cell wall of the baker's yeast, that is, *Saccharomyces cerevisiae* is most abundant in β-glucan. β-Glucan is a water-soluble polysaccharide consists of glucose units. Glucose monomers are linked via β-(1→3) glycosidic bonds in bacteria and algae whereas glucose monomers are linked via β-(1→3) and β-(1→6) glycosidic bonds in yeast and mushrooms. In oats and barley, glucose monomers are linked via β-(1→4) and β-(1→3) glycosidic bonds. β-Glucan obtained from bacteria and algae shows a linear structure whereas β-glucan extracted from yeast, mushrooms, oats, and barley exhibits branched structure. β-Glucan synthesis in cell wall is a complex process because of the identification of large number of different classes of glucans. Several classes of enzymes are involved in synthesis of β-glucan [30]. β-(1→6) Glycosidic side chains interconnect the β-(1→3) glucan chains to create a rigid network [31]. β-Glucan and chitin components are often linked by β-(1→4) linkages. No sharp distinction lies between the insoluble and soluble fractions of β-glucan, however the water solubility of β-glucan is dependent on its structure [32].

Extraction conditions of β-glucan are highly responsible for the ratio of soluble and insoluble fraction [33]. β-Glucans having β-(1→3) linkages and high degree of polymerization are completely insoluble in water which allows interactions between glucan molecules and interaction between glucan molecules and water molecules [34]. Water-solubility and molecular weight of β-glucan is considered to control its hypocholesterolemic effect. It has been reported in literature that viscosity of β-glucan in the gut is mainly responsible for its cholesterol (bad) lowering effects. Viscosity of β-glucan is directly related to its molecular weight, molecular structure, solubility in water, and food matrix [35]. High molecular weight and high solubility in water of β-glucan have high capacity of reducing serum cholesterol as compared to low molecular weight and low soluble β-glucan [36]. This may be due to the higher intestinal viscosity of β-glucan, which reduces the reabsorption of bile acids and leads to higher excretion of bile acids [37]. Higher excretion of bile acids enhances the synthesis of bile acids from cholesterol, which ultimately increase the cholesterol uptake and thus reduces the LDL serum cholesterol. Viscosity of digesta is controlled by β-glucan concentration consumed in the diet and the molecular weight of β-glucan, hence the glycemic response is reported to have significant correlation with concentration and molecular weight of β-glucan [38,39]. β-Glucan with high molecular weight at low concentration forms viscous and pseudoplastic solutions whereas β-glucan with low molecular weight forms softer gel at high concentration [40]. β-1,3-Glucan molecules are almost resistant to the acidic secretions in human stomach. After ingestion, β-glucans gradually passes into the first section of small intestine (duodenum) and are trapped by macrophage receptors located on intestinal wall. These receptors are protein in nature and are produced by bone marrow [41]. When glucan molecule comes in a contact with glucan receptors, it is activated and generates bactericidal compounds, such as lysozyme, reactive oxygen radicals, and oxides. After that cells commence to yield numerous cytokines, which activate the surrounding phagocytes and leukocytes that lead to specific immunity [42].

4.2 Galactomannan

Galactomannan is an endosperm hetero-polysaccharide present in most leguminous seeds and is composed of galactose and mannose [43]. Common sources of galactomannan include guar gum [44], locust bean gum [45], fenugreek, and alfalfa [46]. The galactomannan molecules are resistant to human digestive secretions in small intestine and hence functions as dietary fiber. Even though the difference in sources of extraction, galactomannan shows similar basic structure but shows variations in molecular weight, mannose/galactose ratio, and galactose distribution as side chains on the mannose chain which ultimately affect the physicochemical properties of gum, such as thermal stability, solubility, and viscosity [47].

Guar gum is the most common galactomannan used as dietary fiber source. Guar gum molecule is composed of a linear backbone chain of β-(1$\rightarrow$4) linked mannose units to which galactose units are attached as side chains via α-(1$\rightarrow$6) linkage. Guar gum is obtained from the seed endosperm of *Cyamopsis tetragonolobus* (cluster bean) and generally used as stabilizer and thickener in various foods due to its high viscosity [48]. The ratio of mannose to galactose units has been reported as 2:1 which means galactose units are attached at alternate mannose unit [44]. Native guar galactomannan shows molecular weight of about 900 kDa and viscosity of its 1% solution is about 5500 cps. Whereas partially hydrolyzed guar gum (PHGG), which is used as soluble dietary fiber has low viscosity (4 cps) and low molecular weight of about 8 kDa. PHGG is obtained via partial enzymatic hydrolysis of guar gum at control conditions of pH, temperature, and time. PHGG has similar structure as native guar but with reduced degree of polymerization (DP~29). PHGG when incorporated in the food products do not affect the color, taste, and viscosity of that product.

PHGG is fermented over a longer duration in large intestine due to its medium chain length (DP~29) and provides a better prebiotic effect with the production of high amounts of SCFA when compared to other soluble dietary fibers [49,50]. PHGG is reported to have effects in lowering blood glucose level [51] and serum cholesterol [52]. It is also reported that guar gum reduce the absorption of glucose within rat jejunum due to its gel forming action and by influencing the mucus barrier function [53]. Intake of dietary fiber generally hinders the absorption and utilization of other nutrients whereas PHGG is reported to have no such effects [54,55]. Clinical studies suggest that PHGG is effective in reduction of colonic transit time (CTT), induction of satiety hormone cholecystokinin (CCK) and increased perception of postmeal satiety. The intake of diet incorporated with PHGG slowed the CTT and increased the release of CCK, which may be due to the structure and function of PHGG as it produces huge amounts of SCFA, such as butyrates and propionates that can cause slower CTT and increased release of CCK [56].

4.3 Pectin

Pectin is a type of structural fiber found in the primary cell wall and intracellular layer of plant cells mainly in fruits, such as apples, oranges, lemons, and so on. Citrus fruit contains 0.5%–3.5% pectin which is largely present in peel portion of the fruit. During the ripening process, pectins change to a water-soluble material (ripened fruit) from an insoluble substance (unripe fruit). Pectin is a polymer with linear structure in which few hundred to thousand galacturonic acid monomer units are linked via α-(1→4)-glycosidic bond forming a backbone. The average molecular weight of pectin ranges between 50 and 150 kDa. The backbone of pectin molecule is substituted at certain regions with α-(1→2) rhamnopyranose units from which side chains of galactose, mannose, glucose, and xylose may occur. Methyl esterification of galacturonic acid occurs in pectin. On the basis of methyl esterification, there are two different types of pectin—high methoxyl and low methoxyl pectin. High methoxyl pectins are characterized with more than 50% esterified galacturonic acid residues whereas low methoxyl pectins are characterized with less than 50% esterified galacturonic acid residues [57].

Pectins are reported to have hypocholesterolemic properties but due to its high gelling capacity, it cannot be incorporated in food products at higher concentration as it negatively affects the product sensory characteristics. Pectin is water soluble in nature and bypass the enzymatic digestion process of human small intestine but is easily degraded by the microflora of the colon. In human GI tract, pectin is capable of holding water and forming gel, which ultimately leads to binding of ions and bile acids. Gel forming ability of pectin is considered as possible mechanism of its beneficial health effects, such as improved cholesterol and lipid metabolism, improved gastric emptying, and improved glucose metabolism [58–60]. Pectins are also reported to have some unique abilities for prevention or treatment of diseases, such as intestinal infections, atherosclerosis, cancer, and obesity. In a human study, 24 healthy subjects consumed biscuit fortified with pectin (15 g/day) for 3 weeks. The results of the study suggest that pectin consumption (15 g/day) led to a reduction in total cholesterol by 5% [61]. After the recommendations of Keys et al. (1961), many of the studies have been carried to study the hypocholesterolemic effect of pectin and confirmed the same. In hyperlipidemic men, pectin is associated with a decrease in tensile strength of fibrin and increase in permeability of fibrin. Fibrin is a fibrous, nonglobular protein involved in the clotting of blood. The quality of fibrin is considered as an important risk factor associated with atherosclerosis, stroke, and coronary heart disease. Pectin produces acetate in the human colon, which is assumed to enter in the peripheral circulation and amend the fibrin structure [62]. Pectin is also thought to have a potential role in cancer prevention.

It has been reported in the literature that pectin is capable to bind and decrease tumor growth and cancerous cell migration in rats, which were fed with modified pectin obtained from citrus fruit. The exact mechanism is unclear but it is believed to be a result of galectin inhibitory activity of pectin [63,64]. Pectin performs like a natural prophylactic substance against noxious effect of toxic cations. It is potent in binding and removal of lead and mercury from GI tract [65]. Intravenous injection of pectin reduces the blood coagulation time and control hemorrhage or bleeding. Pectin is also effective in the treatment of diarrheal diseases due to their bactericidal action [57]. Pectin retards the rate of digestion of food components in the intestine via immobilization of food components, which results in low absorption of food. The viscosity of the pectin layer affects the absorption of food components by constraining the contact between the intestinal enzyme and food components [66].

4.4 Psyllium

Psyllium is water-soluble gel-forming mucilage obtained from the seed husk of *Plantago* ovate plant, which is a native herb from regions of Asia, Europe, and North Africa. Mucilage is an adhesive gelatinous substance similar to natural gums. Seed husk of Plantago ovate is a rich source of water-soluble fiber, known as psyllium hydrocolloid or psyllium seed gum. The bioactive component of psyllium is composed of a highly branched arabinoxylan. The main backbone chain consists of xylose units, to which arabinose and xylose are attached as side chains. Arabinoxylans from cereal grains are largely fermented by colonic microflora, however arabinoxylans from psyllium exhibit unrecognized structural characteristic that hinders its fermentation by colonic microflora [67]. Presently, psyllium fortification in various foods, such as ready-to-eat cereals is carried out due to its cholesterol-lowering effects. Several clinical studies in literature reported that psyllium consumption in human reduces 7%–9% serum total cholesterol concentrations [68]. It is also reported that effect of psyllium is independent of fat content and cholesterol content of diet consumed during the clinical study. It is reported that total and

LDL cholesterol (LDL-C) levels are reduced by 0.028 and 0.029 mmol/L by each gram of water-soluble fiber from psyllium. However, no effect on serum HDL cholesterol concentration is reported [58].

Psyllium is reported to exhibit a greater tendency to decrease LDL-C levels and significant improvement in levels of total serum cholesterol. Lowering of total serum cholesterol and LDL-C are consistently observed in the subjects with type II diabetes who consumed psyllium-fortified diet. Exact mechanism of action of psyllium fiber is not clear like other soluble fibers. However, the cholesterol lowering action of psyllium is described by two fundamental hypotheses. According to first hypothesis, psyllium fiber has a tendency to bind or absorb bile acids when they pass through the intestinal lumen and thus prevent their normal reabsorption, which leads to increase in fecal bile acid content and reduce the cholesterol pool. According to second hypothesis, psyllium fiber physically disturbs the intraluminal formation of micelles, which leads to reduction in absorption of cholesterol and reabsorption of bile acids. In both hypotheses, bile acids bound to psyllium fiber are passed to the colon, which leads to higher hepatic conversion of cholesterol into bile acids. This conversion of cholesterol leads to an up-regulation of the LDL receptor and results in higher uptake of LDL-C from the plasma. Overall result is a decrease in serum LDL-C level and hence in total cholesterol level [69]. In 1998, FDA has approved health claim for psyllium that "A food product containing water-soluble fiber from psyllium seed husk, consumed as part of a diet low in saturated fat and cholesterol, may reduce the risk of heart disease."

4.5 Inulin

Inulin is a polymer composed of fructose monomers linked via β-(2-1)-D-frutosyl fructose bonds. It is indigestible in human small intestine due to presence of β-configuration of C-2 and can be fermented by intestinal microflora in large intestine [70]. Nearly 90% of the inulin passes to the colon and digested by colonic bacteria [71]. In chicory inulin, degree of polymerization or the number of monomer unit vary from 2 to 60 showing a combination of both oligomers and polymers [72]. Molecular weight and degree of polymerization (10^3–10^5) of bacterial inulin is very high and more branched as compared to plant inulin [73]. Sources of inulin include onions, garlic, wheat, artichokes, and bananas. Caloric value of inulin is low, that is, 1.5 kcal/g which may be due to its indigestible nature. Inulin is converted to SCFA (e.g., acetate, propionate, and butyrate), lactates and gases [74]. Only the SCFA and lactates are responsible for the caloric value of inulin.

When fermented, inulin has a tendency for propionate production, which leads to decrease in acetate to propionate ratio, which ultimately results in decreased total serum cholesterol and LDL [75].

Inulin is beneficial in reducing the risk of many GI tract diseases, such as irritable bowel diseases and colon cancer. Inulin has also been reported to contribute to the optimal health of the human colon as a prebiotic [76]. It is reported that inulin accelerate the growth of *bifidobacteria* while hampering the growth of potential pathogenic bacteria, such as *Escherichia coli*, *Salmonella*, and *Listeria*. This is helpful in disorders, such as ulcerative colitis and *Clostridium difficile* infections. It is also reported that inulin reduced the biological compounds associated with colonic cancer, including reduced colorectal cell proliferation and water induced necrosis.

Functionality of inulin also includes increased mineral absorption, for example, increased calcium absorption (~20%) was reported in adolescent girls who consumed food supplemented with inulin [77]. Inulin intake is also evidenced with increased bone mineral density when compared to the control subjects. This may be attributed to increased calcium absorption from the colon or an increased solubility in the lumen of the GI tract because of SCFA. It may also enhance mineral absorption through an enrichment of vitamin D. Inulin may have a function in prevention and treatment of obesity. It is reported that inulin increased satiety in adults which led to a decrease in total energy intake. This may be due to the ability of SCFA to increase appetite-suppressing hormones, such as glucagon-like peptide 1 [78].

Inulin is water soluble in nature and is not digested by human digestive enzymes. It produces peculiar results on the effectiveness of the gut, such as reduced pH of intestine; provide support in relieving constipation; and increasing stool volume due to bulking effect. Bulking effect of inulin is similar to bulking effect of other soluble fiber, such as pectin and guar gum [79].

4.6 Resistant Dextrin

Resistant dextrins (RD) are short chain glucose polymers which are strongly resistant to human digestive enzymes and do not have sweet taste [80]. Resistance of RD toward digestive enzymes are due to presence of $(1\rightarrow2)$-, $(1\rightarrow3)$-, $(1\rightarrow6)$-α-, and β-glycoside bonds which is not present in native starch [81]. Dextrinization of starch is observed when starch is heated at high temperature with or without catalyst. Dextrinization involves depolymerization, transglucolyzation, and repolymerization. During the process of dextrinization, isomerization, or formation of new bonds is also observed along with cleavage of bonds. Hydroxyl groups at C-2, C-3, or C-6 glucose unit act on free radicals and undergoes transglycolization, which is based on formation of (1→2)-, (1→3)-, and (1→6)-bonds which leads to formation of branched dextrins [82]. In next phase reversion and recombination reactions occurs which leads to formation of β-(1→6)-glycosidic bonds. RD are associated with all the health benefits caused by dietary fiber, such as PHGG and inulin. Similar to soluble fiber resistant dextrin undergo fermentation in human colon and produces SCFA mainly acetate which accounts for 50% of the SCFA. It is reported that in an in-vitro fermentation study, wheat dextrin produced considerably more total SCFA, propionate and butyrate than PHGG. Hence, RD have higher fermentability by colonic microflora as compared to PHGG [83]. The SCFAs produced in large intestine by soluble fermentable fibers are moderately strongly acidic in nature and hence they reduce the colonic pH and produce acidic environment there. In the large intestine, acidic pH may support the growth of beneficial bacteria, such as bifidobacteria and lactobacilli because they have a strong intrinsic resistance to acid. However, acidic pH prevents the growth of pathogenic bacteria (e.g., *Clostridia*), which is pH sensitive in nature [84]. The formation of SCFAs is associated with better laxation and regularity and also with bulking effect caused by dietary fiber. RD can stimulate pancreatic insulin release via the SCFAs production and affect liver control of glycogen breakdown and ultimately leads to reduction in blood glucose and insulin levels [85]. RD are also associated with reduction in cholesterol levels in humans. The SFCAs produced in large intestine via fermentation of resistant dextrin suppresses the cholesterol synthesis by liver and may reduce LDL-C and triglycerides levels in serum [86].

4.7 Resistant Starch

RS are defined as the starches, which are resistant to the digestion process in human small intestine [87]. RS behave like soluble fiber due to their water solubility and indigestibility without losing their mouth feel and palatability. Hence, resistant starches are important class, which gives benefits of fiber without affecting the sensory characteristics. RS have been classified into four fundamental classes, such as RS1, RS2, RS3, and RS4. RS1 type is physically inaccessible starch composed of starch granules, which are entrapped by indigestible plant material. RS2 type belongs to raw starch granules occur in its natural form, such as in raw potato and high amylose maize. RS3 are retrograded or crystallized starches made by unique cooking and cooling processes of unmodified starch or by food processing operation. RS4 belongs to the starches, which are chemically modified and the modified form becomes resistant to enzymatic digestion. RS are reported to show a reduction in postprandial blood glucose and insulin levels. Among these four classes of resistant starches RS4 type has been reported to have greater glucose lowering effect [88].

It is reported that long-term consumption of the diet containing resistant starch may decrease fasting cholesterol and triglyceride levels. Serum triglyceride levels increases as a result of interactions between sucrose, fructose, and saturated fatty acids. The mechanism behind the reduction of cholesterol level by resistant starches includes increased intestinal viscosity of digesta by resistant starches, which retarded the interactions between sugars and fatty acids [89]. RS increase the intestinal absorption of minerals in rats and humans. It is reported that enhanced intestinal absorption of calcium, magnesium, zinc, iron, and copper in rats observed who were fed with diets rich in resistant starch [90].

5 INSOLUBLE FIBER INTERACTIONS

5.1 Cellulose

Cellulose is an important structural component of primary cell walls in vegetables, green plants, algae, and some bacteria. Cellulose is a linear chain of glucose monomers, which are linked via β-(1→4) glycosidic linkage. It is insoluble in water and resistant to the action of digestive enzymes in the human small intestine and may be fermented by colonic bacteria in the large intestine following production of SCFA. Natural cellulose can be classified into two classes: natural and modified. The crystalline cellulose is made up of intra- and intermolecular noncovalent hydrogen bonds which make it insoluble in water. In last few decades, process have been developed for the preparation of modified celluloses, such as powdered cellulose, microcrystalline cellulose, and hydroxypropylmethyl cellulose which are used as food additives and ingredients. Natural and modified celluloses differ with each other in the amount of crystallization and hydrogen bonding. When these hydrogen bonds are rattled and the crystallinity of cellulose is lost which results in the water solubility of cellulose derivative [91]. Several studies have been reported that studied the influence of cellulose on blood glucose and insulin levels in various models. The results obtained from these studies are extremely contradictory and may be dependent on the subject, cellulose type, and other unexplored factors. Studies using modified cellulose reported more consistent data as compared to studies with natural cellulose. Modified cellulose has been reported to influence lipid metabolism. Hypercholesterolemic adults who were consuming 5 g of Hydroxypropylmethylcellulose per day for a duration of 4 weeks showed a valuable reduction in total and LDL-C [92]. Modified celluloses perform better and more beneficial than native cellulose. These modified celluloses increased the viscosity of the content in GI tract, which is thought to delay nutrient absorption and increase in bile acid excretion.

5.2 Hemicellulose

Hemicelluloses are the polysaccharide and are component of plant cell wall. Hemicelluloses are comprised of different types (hetero) of monomer units. Hemicelluloses show both linear and branched molecules. Hemicelluloses molecules contain 50–200 monomer units and are somewhat smaller than cellulose molecules. Monomer units present in hemicellulose may include pentose monomers (xylose and arabinose) and hexose monomers (glucose, galactose, mannose, rhamnose, glucuronic, and galacturonic acids). As its name indicates, hemicellulose describes a heterogeneous group of chemical structures. Hemicelluloses may exhibit water soluble and insoluble nature [93].

Arabinoxylan is an important type of hemicellulose, which is classified as insoluble dietary fiber. Arabinoxylan is composed of xylose and arabinose monomers. Xylose units are linked via β-1,4-xylosidic linkage to form the backbone to which arabinose units are attached as side chains via α-1,3-linkage. Arabinoxylan forms a major portion of insoluble dietary fiber in whole grains. It forms a major portion of NSP in wheat. It is present in both endosperm and bran portion of the wheat however bran portion is densely loaded with arabinoxylan as compared to endosperm of wheat kernel [94]. Major portion of arabinoxylan is removed as by-product during processing of wheat kernel to wheat flour. In human GI tract, arabinoxylan is immediately fermented by colonic microflora.

Inverse relationship between the levels of intake of arabinoxylan rich bread and postprandial glucose response in healthy adult subjects has been reported [95]. Breads supplemented with arabinoxylan also thought to control blood glucose and insulin in adults having imperfect glucose tolerance [96]. The mechanism of action by which arabinoxylan improve glucose tolerance is not clear. However, a hypothesis is given, according to which high viscosity of the arabinoxylan in the lumen of GI tract reduced the rate of glucose absorption.

5.3 Chitin and Chitosan

Chitin and chitosan are biopolymers, which consist of glucosamine and N-acetylated glucosamine monomer units attached via β-1-4-glycosidic bonds. Sources of chitin and chitosan include shells of arthropods, such as crabs and shrimps [3]. Some fungi and brown algae can also produce chitin and chitosan as exopolysaccharide. Chitin and chitosan differ with each other in terms of acetylation and solubility. Chitin molecule is highly acetylated whereas chitosan molecule is greatly deacetylated. Chitin is insoluble in water and acid whereas chitosan is analogously water-insoluble but soluble in acid. It is reported that chitosan ingestion efficiently reduces serum cholesterol in humans. In a clinical study, chitosan was ingested as biscuits fortified with chitosan by healthy adult males at a dose of up to 6 g/day showed a significant reduction in total serum cholesterol and showed higher levels of serum HDL-cholesterol [97]. Chitosan molecules consisting of more than six units of glucosamine with moderate degree of deacetylation are efficient in inhibiting the cholesterol and lipids absorption in the human intestinal tract. The mechanism behind the reduction in cholesterol absorption is its gel forming ability in the intestinal tract, which binds cholesterol and lipids [98,99]. It is also reported that ascorbic acid enhances gel formation capacity of chitosan and hence increases lipid binding capacity and plasma cholesterol lowering action of chitosan.

A significant unwanted effect of chitosan is the reduction in the absorption of minerals and fat-soluble vitamins, such as vitamin A, D, E, and K. Along with the entrapping lipids and cholesterol; viscous gel of chitosan also binds minerals and fat-soluble vitamins. It is reported that chitosan ingestion results in a symbolic enhancement of fecal excretion of bile acids, such as cholic acid and chemodexoycholic acid in healthy adult males [97]. Chitin may be attributed to disrupt tumor metastasis, as it is a constituent of extracellular matrix.

5.4 Lignin, Suberin, and Cutin

Lignin, suberin, and cutin are complex polymers that occur in cell walls of some specific type of cell. These are present in very small amount in food plants but have significant role in protection against colorectal cancer [100]. Lignin is a highly branched polymer, which is composed of phenylpropanoid units and it is covalently bound to fibrous polysaccharides within plant cell walls. The magnitude of cells having lignified walls often increases with plant maturity and reduces their palatability. Lignin is generally extracted from wood in different ways. Lignin is not a carbohydrate but due to its association with dietary fiber component, it affects the physiological effects of dietary fiber and hence classified as dietary fiber. Suberin is an extracellular biopolymer, which consists of a polyaliphatic domain with a polyaromatic domain, which is obtained from ferulic acid. It occurs in the walls of cork cells that form the skins of many root vegetables and tubers, such as potato tubers. Processed potato peels have been utilized as food additive with enticing baking characteristics in muffins and cookies [101]. Cutin is polyester, which forms the cuticle along with related waxes. This cuticle occurs inside and outside of the outer epidermal wall of leaves and fruits of plants.

Lignin, suberin, and cutin in plant cell walls are considered to safeguard the cell wall polysaccharides from degradation by colonic bacterial enzymes [102]. Plant cell walls become hydrophobic due to the presence of lignin, suberin, and cutin and are potent in vitro adsorbers of hydrophobic carcinogens [103].

REFERENCES

[1] Hispley EH. Dietary fibre and pregnancy toxaemia. Br Med J 1953;2:420–2.
[2] Slavin JL. Dietary fiber: classification, chemical analyses, and food sources. J Am Diet Assoc 1987;87:1164–71.
[3] Champ M, Langkilde AM, Brouns F, Kettlitz B, Collet YLB. Advances in dietary fibre characterisation. Definition of dietary fibre, physiological relevance, health benefits and analytical aspects. Nutr Res Rev 2003;16:71–82.
[4] Ferguson LR, Chavan RR, Harris PJ. Changing concepts of dietary fiber: implications for carcinogenesis. Nutr Cancer 2001;39:155–69.
[5] Tungland BC, Meyer D. Nondigestible oligo and polysaccharides (dietary fibre): their physiology and role in human health and food. Compr Rev Food Sci Food Saf 2002;1:90–109.
[6] Mudgil D, Barak S. Composition, properties and health benefits of indigestible carbohydrate polymers as dietary fiber: a review. Int J Biol Macromol 2013;61:1–6.
[7] Slavin J. Fiber and prebiotics: mechanisms and health benefits. Nutrients 2013;5:1417–35.
[8] Wong JM, Jenkins DJ. Carbohydrate digestibility and metabolic effects. J Nutr 2007;137:2539–46.
[9] Guillon F, Champ M. Structural and physical properties of dietary fibres, and consequences of processing on human physiology. Food Res Int 2000;33:233–45.
[10] Jimenez-Escrig A, Sanchez-Muniz FJ. Dietary fibre from edible seaweeds: chemical structure, physicochemical properties and effects on cholesterol metabolism. Nutr Res 2000;20:585–98.
[11] Slavin JL, Greenberg NA. Partially hydrolyzed guar gum: clinical nutrition uses. Nutrition 2003;19:549–52.
[12] McCarty MF. Nutraceutical resources for diabetes prevention–an update. Med Hypotheses 2005;64:151–8.
[13] Olson A, Gray GM, Chiu M. Chemistry and analysis of soluble dietary fiber. Food Technol 1987;41:71–80.

[14] Langkilde AM, Andersson H. Amount and composition of substrate entering the colon. In: Guillon F, Amadó R, Amara-Collaco MT, Andersson H, Asp NG, Bach Knudsen KE, Champ M, Mathers J, Robertson JA, Rowland I, Van Loo J, editors. Functional properties of nondigestible carbohydrates. Nantes: INRA; 1998. p. 140–2.

[15] Elleuch M, Bedigian D, Roiseux O, Besbes S, Blecker C, Attia H. Dietary fibre and fibre-rich by-products of food processing: characterisation, technological functionality and commercial applications: a review. Food Chem 2011;124:411–21.

[16] Edwards C. Dietary fibre, fermentation and the colon. In: Cherbut C, Barry JL, Lairon D, Durand M, editors. Dietary fibre: mechanisms of action in human physiology and metabolism. Paris: John Libbey Eurotex; 1995. p. 51–60.

[17] Bingham SA. Mechanisms and experimental and epidemiological evidence relating dietary fibre (non-starch polysaccharides) and starch to protection against large bowel cancer. Proc Nutr Soc 1990;49:153–71.

[18] Topping DL, Pant I. Short-chain fatty acids and hepatic lipid metabolism: experimental studies. In: Cummings JH, Rombeau JC, Sakata T, editors. Physiological and clinical aspects of short-chain fatty acids. Cambridge: Cambridge University Press; 1995. p. 495–508.

[19] Salminen S, Bouley C, Boutron-Ruault MC, Cummings JH, Franck A, Gibson GR, et al. Functional food science and gastrointestinal physiology and function. Br J Nutr 1998;80:147–71.

[20] Macfarlane GT, Macfarlane S. Factors affecting fermentation reactions in the large bowel. Proc Nutr Soc 1993;52:367–73.

[21] Dikeman CL, Fahey GC Jr. Viscosity as related to dietary fiber: a review. Crit Rev Food Sci Nutr 2006;46:649–63.

[22] Morris ER. Shear thinning of random coil polysaccharides: characterisation by two parameters from a simple linear plot. Carbohyd Polym 1990;13:85–96.

[23] Morris ER. Physico-chemical properties of food polysaccharides. In: Schweizer TF, Edwards C, editors. Dietary fibre—a component of food: nutritional function in health and disease. Berlin: Springer-Verlag, ILSI Europe; 1992. p. 41–56.

[24] Fleury N, Lahaye M. Chemical and physico-chemical characterisation of fibres from *Laminaria digitata* (kombu breton): a physiological approach. J Sci Food and Agric 1991;55:389–400.

[25] Thibault JF, Lahaye M, Guillon F. Physicochemical properties of food plant cell walls. In: Schweizer TF, Edwards C, editors. Dietary fibre, a component of food. nutritional function in health and disease. Berlin: Springer-Verlag, ILSI Europe; 1992. p. 21–39.

[26] Thibault JF, Renard MGC, Guillon F. Physical and chemical analysis of dietary fibres in sugar-beet and vegetable. In: Jackson JF, Linskens HF, editors. Modern methods of plant analysis, Vol. XVI. Heidelberg: Springer-Verlag; 1994. p. 23–55.

[27] Robertson JA. Application of plant-based byproducts as fiber supplements in processed foods. Recent Res Dev Agric Food Chem 1998;2:705–17.

[28] Cherbut C. Fermentation et fonction digestive colique. Cahiers de Nutrition et de Diététique 1995;30:143–7.

[29] Du B, Bian Z, Xu B. Skin health promotion effects of natural beta-glucan derived from cereals and microorganisms: a review. Phytother Res 2014;28:159–66.

[30] Douglas CM. Fungal β-(1, 3)-D-glucan synthesis. Med Mycol 2001;39:55–66.

[31] Bowman SM, Free SJ. The structure and synthesis of the fungal cell wall. Bioessays 2006;28:799–808.

[32] Havrlentova M, Petrulakova Z, Burgarova A, Gago F, Hlinkova A, Sturdik E. Cereal β-glucans and their significance for the preparation of functional foods-a review. Czech J Food Sci 2011;29:1–14.

[33] Virkki L, Johansson L, Ylinen M, Maunu S, Ekholm P. Structural characterization of water-insoluble nonstarchy polysaccharides of oats and barley. Carbohyd Polym 2005;59:357–66.

[34] Bohn JA, BeMiller JN. (1→ 3)-β-D-Glucans as biological response modifiers: a review of structure–functional activity relationships. Carbohyd Polym 1995;28:3–14.
[35] Bae I, Lee S, Kim S, Lee H. Effect of partially hydrolyzed oat β-glucan on the weight gain and lipid profile of mice. Food Hydrocoll 2009;23:2016–21.
[36] Kerckhoffs D, Hornstra G, Mensink R. Cholesterol-lowering effect of beta-glucan from oat bran in mildly hypercholesterolemic subjects may decrease when beta-glucan is incorporated into bread and cookies. Am J Clin Nutr 2003;78:221–7.
[37] Andersson K, Svedberg K, Lindholmb M, Ostec R, Hellstranda P. Oats (*Avena sativa*) reduce atherogenesis in LDL-receptor-deficient mice. Atherosclerosis 2010;212:93–9.
[38] Wood PJ, Beer MU, Butler G. Evaluation of the role of concentration and molecular weight of oat β-glucan in determining effect of viscosity on plasma glucose and insulin following an oral glucose load. Br J Nutr 2000;84:19–23.
[39] Wood PJ. Cereal β-glucans in diet and health. J Cereal Sci 2007;46:230–8.
[40] Doublier JL, Wood PJ. Rheological properties of aqueous solutions of (1-3) (1-4)- β-D-glucan from oats (*Avena sativa L.*). Cereal Chem 1995;72:335–40.
[41] Duckova K, Bukovsky M, Kucera J. Study of topical dispersions with an immunomodulatory activity. STP Pharma Sci 1997;7:223–8.
[42] Okamoto T, Kodoi R, Nonaka Y, Fukuda I, Hashimoto T, Kanazawa K, et al. Lentinan from shiitake mushroom (*Lentinus edodes*) suppresses expression of cytochrome P450 1A subfamily in the mouse liver. Biofactors 2004;21:407–9.
[43] Sandolo C, Matricardi P, Alhaique F, Coviello T. Dynamo-mechanical and rheological characterization of guar gum hydrogels. Eur Polym J 2007;43:3355–67.
[44] Mudgil D, Barak S, Khatkar BS. Guar gum: processing, properties and food applications: a review. J Food Sci Technol 2014;51:409–18.
[45] Barak S, Mudgil D. Locust bean gum: Processing, properties and food applications—a review. Int J Biol Macromol 2014;66:74–80.
[46] Whistler RL, Hymowitz T. Guar: agronomy, production, industrial use and nutrition. West Lafayette: Purdue University Press; 1979.
[47] Srivastava M, Kapoor VP. Seed galactomannans: an overview. Chem Biodiv 2005;2: 295–317.
[48] Mudgil D, Barak S, Khatkar BS. Soluble fibre and cookie quality. Agro Food Ind Hi Tech 2012;23:15–7.
[49] Ohashi Y, Harada K, Tokunaga M, Ishihara N, Okubo T, Ogasawara Y, et al. Fecal fermentation of partially hydrolyzed guar gum. J Funct Foods 2012;4:398–402.
[50] Pylkas AM, Juneja LR, Slaving JL. Comparison of different fibers for in vitro production of short chain fatty acids by intestinal microflora. J Med Food 2005;8:113–6.
[51] Gu Y, Yamashita T, Suzuki I, Juneja LR, Yokawa T. Effect of enzyme hydrolyzed guar gum on the elevation of blood glucose levels after meal. Med Biol 2003;147:19–24.
[52] Kajimoto O, Xiao J, Kondo S, Iwatsuki K, Kokubo S, Sakamoto A, et al. Suppression of postprandial serum triglyceride elevation by a drinking yogurt supplemented with partially hydrolyzed guar gum. J Nutr Food 2004;7:1–17.
[53] Johnson IT, Gee JM. Effect of gel-forming gums on the intestinal unstirred layer and sugar transport in vitro. Gut 1981;22:398–403.
[54] Takahashi H, Yang SI, Hayashi C, Kim M, Yamanaka J, Yamamoto T. Effect of partially hydrolyzed guar gum on fecal output in human volunteers. Nutr Res 1993;13: 649–57.
[55] Alam NH, Meier R, Rausch T, Meyer-Wyss B, Hildebrand P, Schneider H, et al. Effects of a partially hydrolyzed guar gum on intestinal absorption of carbohydrate, protein and fat: a double-blind controlled study in volunteers. Clin Nutr 1998;17:125–9.
[56] Rao TP. Role of guar fiber in appetite control. Physiol Behav 2016;164:277–83.
[57] Thakur BR, Singh RK, Handa AK. Chemistry and uses of pectin—a review. Crit Rev Food Sci Nutr 1997;37:47–73.

[58] Brown L, Rosner B, Willett WW, Sacks FM. Cholesterol-lowering effects of dietary fiber: a meta-analysis. Am J Clin Nutr 1999;69:30–42.
[59] Lawaetz O, Blackburn AM, Bloom SR, Aritas Y, Ralphs DNL. Effect of pectin on gastric-emptying and gut hormone-release in the dumping syndrome. Scand J Gastroenterol 1983;18:327–36.
[60] Jenkins DJA, Leeds AR, Gassull MA, Cochet B, Alberti KGM. Decrease in postprandial insulin and glucose concentrations by guar and pectin. Ann Intern Med 1977;86: 20–3.
[61] Keys A, Grande F, Anderson JT. Fiber and pectin in the diet and serum cholesterol concentration in man. Exp Biol Med 1961;106:555–8.
[62] Veldman FJ, Nair CH, Vorster HH, Vermaak WJ, Jerling JC, Oosthuizen W, et al. Possible mechanisms through which dietary pectin influences fibrin network architecture in hypercholesterolaemic subjects. Thromb Res 1999;93:253–64.
[63] Kolatsi-Joannou M, Price KL, Winyard PJ, Long DA. Modified citrus pectin reduces galectin-3 expression and disease severity in experimental acute kidney injury. PLoS One 2011;6:1–9.
[64] Nangia-Makker P, Hogan V, Honjo Y, Baccarini S, Tait L, Bresalier R, et al. Inhibition of human cancer cell growth and metastasis in nude mice by oral intake of modified citrus pectin. J Natl Cancer Inst 2002;94:1854–62.
[65] Kohn R. Binding of toxic cations to pectin, its oligomeric fragment and plant tissues. Carbohyd Polym 1982;2:273–5.
[66] Dunaif G, Schneeman BO. The effect of dietary fibre on human pancreatic enzyme activity in vitro. Am J Clin Nutr 1981;34:1034–5.
[67] Marlett JA, Fischer MH. The active fraction of psyllium seed husk. Proc Nutr Soc 2003;62:207–9.
[68] Garvin JE, Forman DT, Eiseman WR, Phillips CR. Lowering of human serum cholesterol by an oral hydrophilic colloid. Exp Biol Med 1965;120:744–6.
[69] Rodrıguez-Moran M, Guerrero-Romero F, Lazcano-Burciaga G. Lipid- and glucose-lowering efficacy of *Plantago psyllium* in type II diabetes. J Diabetes Complicat 1998;12:273–8.
[70] Apolinario AC, de Lima Damasceno BP, de Macedo Beltrao NE, Pessoa A, Converti A, da Silva JA. Inulin-type fructans: a review on different aspects of biochemical and pharmaceutical technology. Carbohyd Polym 2014;101:368–78.
[71] Cherbut C. Inulin and oligofructose in the dietary fibre concept. Br J Nutr 2002;87: 159–62.
[72] Roberfroid MB. Introducing inulin-type fructans. Br J Nutr 2005;93:13–25.
[73] Cho SS, Samuel P. Fiber ingredients: food applications and health benefits. Boca Raton, FL, USA: CRC Press; 2009. p. 41–55.
[74] Nyman M. Fermentation and bulking capacity of indigestiblecarbohydrates: the case of inulin and oligofructose. Br J Nutr 2002;87:163–8.
[75] Amaral L, Morgan D, Stephen AM, Whiting S. Effect of propionate on lipid-metabolism in healthy-human subjects. FASEB J 1992;6:1655–6.
[76] Gibson GR, Beatty ER, Wang X, Cummings JH. Selective stimulation of bifidobacteria in the human colon by oligofructose and inulin. Gastroenterology 1995;108: 975–82.
[77] Griffin IJ, Hicks PMD, Heaney RP, Abrams SA. Enriched chicory inulin increases calcium absorption mainly in girls with lower calcium absorption. Nutr Res 2003;23: 901–9.
[78] Cani PD, Joly E, Horsmans Y, Delzenne NM. Oligofructose promotes satiety in healthy human: a pilot study. Eur J Clin Nutr 2006;60:567–72.
[79] Anderson JW, Baird P, Davis RH, Ferreri S, Knudtson M, Koraym A, et al. Health benefits of dietary fiber. Nutr Rev 2009;67:188–205.

[80] Ohkuma K, Matsuda I, Katta Y, Hanno Y. Pyrolysis of starch and its digestibility by enzymes—characterization of indegestible dextrin. Denpun Kagaku 1999;37:107–14.

[81] Mana KK. Enzyme resistant dextrins from high amylose corn mutant starches. Starch/ Stärke 2003;53:21–6.

[82] Tomasik P, Wiejak S, Pałasinski M. The thermal decomposition of carbohydrates. Part II. The decomposition of starch. Adv Carbohyd Chem Biochem 1989;47: 279–344.

[83] Stewart ML, Savarino V, Slavin JL. Assessment of dietary fiber fermentation: effect of *Lactobacillus reuteri* and reproducibility of shortchain fatty acid concentrations. Mol Nutr Food Res 2009;53:114–20.

[84] Topping DL, Clifton PM. Short-chain fatty acids and human colonic function: roles of resistant starch and nonstarch polysaccharides. Physiol Rev 2001;81:1031–64.

[85] Wolever TM, Jenkins DJ. Effect of dietary fiber and foods on carbohydrate metabolism. In: Spiller GA, editor. CRC handbook of dietary fiber in human nutrition. 2nd ed. Boca Raton: CRC Press; 1993. p. 111–52.

[86] Ohkuma K, Wakabayashi S. Fibersol-2: A soluble, non-digestible, starch derived dietary fibre. In: McCleary B, Prosky L, editors. Advanced dietary fibre technology. Oxford: Blackwell Science; 2001. p. 510–23.

[87] Higgins JA. Resistant starch: metabolic effects and potential health benefits. J AOAC Int 2004;87:761–8.

[88] Haub MD, Hubach KL, Al-Tamimi EK, Ornelas S, Seib PA. Different types of resistant starch elicit different glucose reponses in humans. J Nutr Metab 2010;2010:1–4.

[89] Porikos KP, Vanitallie TB. Diet-induced changes in serum transaminase and triglyceride levels in healthy adult men—role of sucrose and excess calories. Am J Med 1983;75:624–30.

[90] Lopez HW, Levrat-Verny MA, Coudray C, Besson C, Krespine V, Messager A, et al. Class 2 resistant starches lower plasma and liver lipids and improve mineral retention in rats. J Nutr 2001;131:1283–9.

[91] Takahashi R, Hirasawa Y, Nichinari K. Cellulose and its derivatives. Foods Food Ingredients J Japan 2003;208:824–8.

[92] Maki KC, Carson ML, Anderson WHK, Geohas J, Reeves MS, Farmer MV, et al. Lipid-altering effects of different formulations of hydroxypropylmethylcellulose. J ClinLipidol 2009;3:159–66.

[93] Hu G, Huang S, Cao S, Ma Z. Effect of enrichment with hemicellulose from rice bran on chemical and functional properties of bread. Food Chem 2009;115:839–42.

[94] Ring SG, Selvendran RR. Isolation and analysis of cell-wall material from beeswing wheat bran (*Triticum-aestivum*). Phytochemistry 1980;19:1723–30.

[95] Lu ZX, Walker KZ, Muir JG, Mascara T, O'Dea K. Arabinoxylan fiber, a byproduct of wheat flour processing, reduces the postprandial glucose response in normoglycemic subjects. Am J Clin Nutr 2000;71:1123–8.

[96] Lu ZX, Walker KZ, Muir JG, O'Dea K. Arabinoxylan fibre improves metabolic control in people with Type II diabetes. Eur J Clin Nutr 2004;58:621–8.

[97] Maezaki Y, Tsuji K, Nakagawa Y, Kawai Y, Akimoto M, Tsugita T, et al. Hypocholesterolemic effect of chitosan in adult males. Biosci Biotechnol Biochem 1993;57:1439–44.

[98] Deuchi K, Kanauchi O, Shizukuishi M, Kobayashi E. Continuous and massive intake of chitosan affects mineral and fat-soluble vitamin status in rats fed on a high-fat diet. Biosci Biotechnol Biochem 1995;59:1211–6.

[99] Deuchi K, Kanauchi O, Imasato Y, Kobayashi E. Effect of the viscosity or deacetylation degree of chitosan on fecal fat excreted from rats fed on a high-fat diet. Biosci Biotechnol Biochem 1995;59:781–5.

[100] Ferguson LR, Harris PJ. Suberized plant cell walls suppress formation of heterocyclic amine-induced aberrant crypts in a rat model. Chem-Biol Inter 1998;114:191–209.

[101] Arora A, Camire ME. Performance of potato peels in muffins and cookies. Food Res Int 1994;27:15–22.
[102] Harris PJ, Ferguson LR. Dietary fibre: its composition and role in protection against colorectal cancer. Mutat Res 1993;290:97–110.
[103] Harris PJ, Triggs CM, Roberton AM, Watson ME, Ferguson LR. The adsorption of heterocyclic aromatic amines by model dietary fibres with contrasting compositions. Chem Biol Interact 1996;100:13–25.

CHAPTER 4

The Influence of Fiber on Gut Microbiota: Butyrate as Molecular Player Involved in the Beneficial Interplay Between Dietary Fiber and Cardiovascular Health

Lorella Paparo, Antonio Calignano, Carlo G. Tocchetti, Carmen Di Scala, Roberto Russo, Domenico Bonaduce, Roberto B. Canani

1 INTRODUCTION

Cardiovascular diseases (CVD) are the main cause of death in industrialized countries and are responsible for high economic health care costs [1]. The World Health Organization estimated about 20 million CVD deaths in 2015, rendering CVD as the major contributor to mortality worldwide and a significant proportion of the population have risk factors that contribute to the development of CVD [2].

Diet has a pivotal role in the development and prevention of CVD, and it is considered an important target of intervention. A diet low in saturated fats, with plenty of fiber, significantly reduces the risk of major cardiac events [3].

2 DIETARY FIBERS SHAPE GUT MICROBIOTA COMPOSITION AND FUNCTION

Dietary fibers are complex carbohydrates consisting of both soluble and insoluble components. Dietary guidelines recommend that women and men should respectively eat around 25 and 38 g of fiber per day, but American adults eat just 15 g daily on average [4]. It is now clear that dietary fiber does not just feed humans, but it also feeds the trillions of microbes in our gut, which influence microbiota composition and function. The gut microbiota numbers over 10^{14} microbial cells that collectively contain at least 150 times more genes than their human host [5]. In recent years, with the advancement in high-throughput molecular sequencing, great strides

Dietary Fiber for the Prevention of Cardiovascular Disease
http://dx.doi.org/10.1016/B978-0-12-805130-6.00004-5

have been taken up not only to determine the diversity, abundance, and alteration–dysbiosis of gut microbiota, but also to uncover its functions and therapeutic potential. Because of these new technologies in animal models and in clinical studies, a correlation between gut microbiota composition and CVD onset has been demonstrated. Studies have also shown that targeting intestinal microbiota via dietary treatment ameliorates metabolic distortions implicated in CVD risk [6]. Also, individuals living for 1 month on a low-fiber diet present significant modification of the complex microbial community that we harbor in our intestinal tract (dysbiosis) [7]. On the contrary, the consumption of fibers, fruit, and vegetables is associated with a beneficial effect on gut microbiota with increased microbial richness, at either taxonomic or gene level [8–10], a condition that is considered to be protective against many pathologic conditions. Epidemiological studies and intervention trials have demonstrated the protective effects associated with a Mediterranean diet (MD) in prevention of CVD [11,12]. Subjects consuming a high-level of cereals, fruits, vegetables, and legumes, consistent with the MD, show a higher number of fiber-degrading bacteria in the gut microbiota (i.e., *Firmicutes* and *Bacteroidetes*) and associated increased fecal short-chain fatty acid (SCFAs) levels (Fig. 4.1). In contrast, the lower adherence to the MD corresponded to an increase of urinary trimethylamine oxide (TMAO) levels. We have previously demonstrated a significant posi-

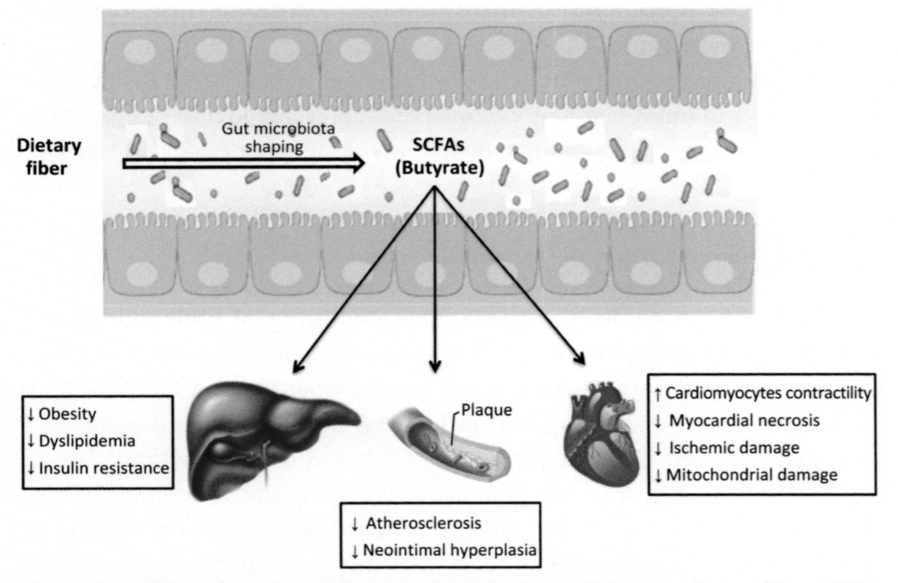

Figure 4.1 ***Pathways and cardiovascular health benefits driven by dietary fiber.*** *SCFAs*, Short-chain fatty acids.

tive correlation between urinary TMAO and a number of *L.ruminococcus* bacteria [13]. Products from a protein-enriched diet, such as L-carnitine and phosphatidylcholine, may be metabolized and converted to trimethylamine (TMA) [14]. Consequently, plasma levels of choline, betaine, and TMA are strongly correlated with the progression of atherosclerosis and increased CVD risk [15]. Plasma choline levels are strong predictors of major cardiac events in patients with suspected acute coronary syndrome [16], as well as in individuals with stable coronary disease.

It has been shown that bacteria from the class *Erysipelotrichia (*phylum *Firmicutes*) may promote cholesterol accumulation and the formation of atherosclerotic plaque [17]. This bacterial species is able to metabolize carnitine to TMAO [18]. A clinical study has demonstrated that TMAO is also a predictor of incident death, myocardial infarction, and stroke, as well as CVD even in individuals at low risk of cardiovascular events [19]. It has also been observed that a *Prevotella* enterotype correlates with higher TMAO levels [20].

Dysbiosis of the gut microbiota has been also reported for several conditions, especially for metabolic diseases [21]. Metabolic diseases are well-known common risk factors for development of severe CVD [22], such as atherosclerosis, stroke, and myocardial infarction.

Diet-induced gut microbiota dysbiosis may result in bacterial translocation into the systemic blood flow, characterized by a plaque-specific blood-like microbiota dominated by *Proteobacteria* phylum [23]. Whole-genome sequencing demonstrated an increased abundance of the genus *Collinsella* and a decrease of *Roseburia* and *Eubacterium* in stool samples from patients with atherosclerosis compared to healthy controls. In addition, a negative correlation with C-reactive protein (a biomarker associated with increased risk of CVD) and Clostridiales genera *Clostridium* and *Peptostreptococcus* has been shown. [24].

Intriguingly, congenital heart disease, the first cause of premature death in the first year of life, is also associated with gut microbiota dysbiosis. A study demonstrated that infants with congenital heart disease have a lower level of total *Bacteroidetes* and *Bifidobacteria* compared to healthy controls [25].

3 BUTYRATE IS THE "CURRENCY" ADOPTED BY THE GUT MICROBIOTA THAT INFLUENCES CARDIOVASCULAR HEALTH

SCFAs, such as acetate, propionate, and butyrate, are produced in millimolar concentrations (approximately 60:25:15 mM, respectively) by anaerobic fermentation of dietary fiber by commensal gut bacteria [26]. Among the SCFAs, butyrate has received particular attention for its multiple beneficial effects at the intestinal and extraintestinal level [26,27]. Butyrate-producing bacteria represent a functional group, rather than a coherent phylogenetic group. Two of the most important groups of butyrate producers appear to be *Faecalibacterium prausnitzii*, which belongs to the *Clostridium leptum* cluster (or clostridial cluster IV), and *Eubacterium rectale/Roseburia intestinalis* spp., which belong to the *Clostridium coccoides* cluster (or clostridial cluster XIVa) within the *Firmicutes* taxa [28]. *Faecalibacterium prausnitzii* and *Roseburia intestinalis* are a prototypical antiinflammatory components of the gut microbiota and butyrate producers. A low concentration of SCFAs-producing Clostridiales has been demonstrated in CVD patients [29], and in several CVD-associated metabolic conditions, such as metabolic syndrome, diabetes, and obesity [30–34]. The butyrate concentration in the colon is highly variable, ranging from 5 to 30 mM [35]. The gut butyrate concentration is the result from the balance between butyrate biosynthesis by the gut microbiota, absorption, and utilization primarily via β-oxidation in the mitochondria of intestinal epithelial cells, and excretion. Upon absorption, the vast majority of butyrate is metabolized locally, while an amount of butyrate enters into the bloodstream. The concentration range of butyrate in the systemic circulation is 0.1–1 mM under normal physiologic conditions [26]. Up to 95% of total butyrate is either absorbed via passive diffusion through ionic exchanges or actively transported across the gut epithelium. Monocarboxylate transporter 1 and sodium-coupled monocarboxylate transporter 1 (SLC5A8) have been identified [36] as butyrate transporters across the plasma membrane. Butyrate also binds G protein-coupled receptors (GPR, or free fat acid receptor), particularly GPR41, GPR43, and GPR109A, on surface of several cell types [37–41]. These receptors are coupled to Gαi and/or Gαq. As a result, activation of these receptors leads to suppression of adenylate cyclase/cAMP signaling pathway and activation of phospholipase C and increase in intracellular calcium levels [42,43].

It has been demonstrated that butyrate is a peroxisome proliferator-activated receptor (PPAR) γ agonist [44,45]. PPARs are widely present in platelet and play an important role in platelet aggregation [46–48]. Our preliminary experimental data support this hypothesis. Indeed, human platelets treated with sodium butyrate at the same plasma concentrations found in healthy humans, showed a reduced aggregation after collagen or ADP stimulation (data not published). Classical drugs PPAR-α agonist, such as statins, are able to stimulate significant antiplatelet and antithrombotic activity [46–49]. In addition, PPAR receptors are expressed on vascular endothelium and vascular smooth muscle suggesting their involvement in the regulation of vascular tone [50]. It has also been demonstrated that PPAR-α agonists, due to their antiinflammatory properties, could reduce myocardial inflammation and fibrosis induced by high level of Angiotensin II in the rat [51].

In addition, butyrate can exert its biological effects by diverse molecular and epigenetic mechanisms. Epigenetics is the study of mitotically heritable, yet potentially reversible, molecular modifications to DNA, and chromatin without alterations to the underlying DNA sequence [52]. Butyrate has been shown to be a potent inhibitor of histone deacetylase (HDACs) activity, causing changes in the acetylation status of chromatin and other nonhistone proteins and resulting in epigenetic modulation of gene expression leading to different cellular effects [53]. Moreover, butyrate can also contribute to decreased DNA methylation through the inhibition of DNA methyltransferase 3A (DNMT3A) expression, the enzyme responsible for de novo DNA methylation [54]. These epigenetic effects of butyrate promote an "opened" chromatin conformation that contributes to regulating gene expression. Recent RNA-seq analysis reveals that butyrate significantly alters the expression of 11,443 genes in vitro, representing ~65% of all genes transcribed in the epithelial transcriptome [55].

Interestingly, HDAC inhibition has been demonstrated to play a key role in protecting the heart from pathologic hypertrophy and ischemia [56–60] and several observations have implicated HDACs inhibitors, such as butyrate, in the control of cardiac hypertrophy. Indeed, treatment of rat cardiomyocytes with HDAC inhibitors prevented hypertrophy, sarcomere organization, and activation of the fetal gene program normally evoked by hypertrophic agonists, suggesting that these inhibitors fully antagonize the hypertrophic program rather than selectively inhibit only a subset of genes involved in this process [56]. The HDAC inhibitor, TSA, increased resistance to myocardial ischemia and reperfusion injury, as indicated by an improvement in the recovery of ventricular function and the reduction of myocardial necrosis when mice were injected with TSA [60], and changes in gut microbiota composition and serum butyrate levels have also been correlated with neointimal hyperplasia development after carotid artery balloon angioplasty in rats following treatment with oral vancomycin and dietary supplementation of sodium butyrate. Cavasin et al. have demonstrated that HDAC inhibitors are able to reduce right ventricular hypertrophy and to exert beneficial effects on the right ventricular, including suppression of pathological gene expression, inhibition of proapoptotic caspase activity, and repression of proinflammatory protein expression [61,62]. Moreover, recently it has been pointed out that HDAC inhibitors induce an early differentiation of megakaryocytic differentiation in to blood cells lines and platelet production indicating an involvement of HADAC modulation in the hematopoietic process [16]. Vancomycin mediated neointimal hyperplasia after angioplasty and lowered serum butyrate concentration in rats, could potentially be reversed by butyrate supplementation [63]. Several studies show that DNA demethylation is associated with the development of atherosclerosis [64–66]. Yideng et al. demonstrated that the expression of PPARα,γ can be inhibited by homocysteine, implicated as a common and independent risk factor for atherosclerosis, through promoter hypomethylation in human monocytes [66].

Another epigenetic mechanism regulated by butyrate is DNA demethylation, through the inhibition of DNMT [54]. Treatment with the DNMT inhibitor can upregulate SOD2 expression. This gene has an important role in the development of atherosclerosis plaques because it activates HIF-1α [67]. Doxorubicin, a natural anthracycline, is a potent antineoplastic agent that is being used as a frontline cancer chemotherapeutic [68], but this drug can also cause a well-known cardiomyopathy [69]. Rephaeli et al. showed that a prodrug of butyric acid in combination with doxorubicin is able to prevent the damage to cultured cardiomyocytes, caused by doxorubicin alone [70]. In vivo, the cardioprotective effect of butyrate against doxorubicin toxicity was associated with a reduction in proinflammatory factors in the heart and modulation of expression of proteins involved in cytoprotection and angiogenesis [71]. Data on mouse and rats model demonstrated that cultured cardiomyocytes and rodent hearts display reduced ischemia damage when treated with a prodrug of butyric acid and that upregulation of the cardioprotective enzyme HO-1 concurs with the protection attained. Moreover, treatment with butyrate was able to reduce the mitochondrial damage and cardiomyocytes death [72].

4 CONCLUSIONS

An "unhealthy" diet modulates negatively the gut microbiota and its metabolism, leading to development of several diseases. Because of the novel technologies to investigate gut microbiota composition and function, as well as the huge amount of data on the influence of gut microbiota on the occurrence of CVD and CVD-associated metabolic conditions, we are discovering the crucial pathways of the beneficial effects exerted by dietary fiber on these conditions. In this light, several beneficial effects elicited by the Mediterranean diet on CVD risk could be related to a positive gut microbiota composition leading to a subsequent increased butyrate production, which in turn could be responsible for many extra-intestinal effects. All these findings are reinforcing the actual dietary strategies against CVD and are paving the way for future possible therapeutic interventions involving gut microbiota-derived compounds, such as butyrate.

5 ABBREVIATIONS

CVD Cardiovascular diseases
GPR G Protein-coupled receptors
HADC Inhibitor of histone deacetylase
MD Mediterranean diet
PPAR Peroxisome proliferator-activated receptor
SCFA Short-chain fatty acid
TMA Trimethylamine
TMAO Trimethylamine *N*-oxide

REFERENCES

[1] Graham I, Atar D, Borch-Johnsen K, Boysen G, Burell G, Cifkova R, et al. European guidelines on cardiovascular disease prevention in clinical practice: full text. Fourth Joint Task Force of the European Society of Cardiology and other societies on cardiovascular disease prevention in clinical practice (constituted by representatives of nine societies and by invited experts). Eur J Cardiovasc Prev Rehabil 2007;14(Suppl. 2):S1–113.
[2] Khan M, Mensah GA. Epidemiology of cardiovascular disease. In: Fuster V, Kelly BB, editors. Promoting cardiovascular health in the developing world a critical challenge to achieve global health. Washington, DC: National Academies Press; 2010.
[3] Thorburn AN, Macia L, Mackay CR. Diet, metabolites, and "western-lifestyle" inflammatory diseases. Immunity 2014;40(6):833–42.
[4] King DE, Mainous AG, Lambourne CA. Trends in dietary fiber intake in the United States, 1999–2008. J Acad Nutr Diet 2012;112(5):642–8.
[5] Clemente JC, Ursell LK, Parfrey LW, Knight R. The impact of the gut microbiota on human health: an integrative view. Cell 2012;148(6):1258–70.
[6] Hansen TH, Gøbel RJ, Hansen T, Pedersen O. The gut microbiome in cardio-metabolic health. Genome Med 2015;7(1):33.
[7] Pendyala S, Walker JM, Holt PR. A high-fat diet is associated with endotoxemia that originates from the gut. Gastroenterology 2012;142(5). 1100-1.
[8] Claesson MJ, Jeffery IB, Conde S, et al. Gut microbiota composition correlates with diet and health in the elderly. Nature 2012;488:178–84.
[9] Cotillard A, Kennedy SP, Kong LC, et al. Dietary intervention impact on gut microbial gene richness. Nature 2013;500:585–8.
[10] Le Chatelier E, Nielsen T, Qin J, et al. Richness of human gut microbiome correlates with metabolic markers. Nature 2013;500:541–6.
[11] Estruch R, Ros E, Salas-Salvado J, et al. Primary prevention of cardiovascular disease with a Mediterranean diet. N Engl J Med 2013;368:1279–90.
[12] de Lorgeril M, Renaud S, Mamelle N, et al. Mediterranean alphalinolenic acid-rich diet in secondary prevention of coronary heart disease. Lancet 1994;343:1454–9.

[13] De Filippis F, Pellegrini N, Vannini L, Jeffery IB, La Storia A, Laghi L, et al. High-level adherence to a Mediterranean diet beneficially impacts the gut microbiota and associated metabolome. Gut 2015;65:1812–21.
[14] Serino M, Blasco-Baque V, Nicolas S, Burcelin R. Far from the eyes, close to the heart: dysbiosis of gut microbiota and cardiovascular consequences. Curr Cardiol Rep 2014;16:540.
[15] Wang Z, Klipfell E, Bennett BJ, et al. Gut flora metabolism of phosphatidylcholine promotes cardiovascular disease. Nature 2011;472:57–63.
[16] Danne O, Lueders C, Storm C, Frei U, Möckel M. Whole blood choline and plasma choline in acute coronary syndromes: prognostic and pathophysiological implications. Clin Chim Acta 2007;383:103–9.
[17] Goldsmith JR, Sartor RB. The role of diet on intestinal microbiota metabolism: downstream impacts on host immune function and health, and therapeutic implications. J Gastroenterol 2014;49(5):785–98.
[18] Spencer MD, Hamp TJ, Reid RW, Fischer LM, Zeisel SH, Fodor AA. Association between composition of the human gastrointestinal microbiome and development of fatty liver with choline deficiency. Gastroenterology 2011;140(3):976–86.
[19] Tang WHW, Wang Z, Levison BS, Koeth RA, Britt EB, Fu X, et al. Intestinal microbial metabolism of phosphatidylcholine and cardiovascular risk. N Engl J Med 2013;368:1575–84.
[20] Koeth RA, Wang Z, Levison BS, Buffa JA, Org E, Sheehy BT, et al. Intestinal microbiota metabolism of L-carnitine, a nutrient in red meat, promotes atherosclerosis. Nat Med 2013;19(5):576–85.
[21] Serino M, Luche E, Chabo C, Amar J, Burcelin R. Intestinal microflora and metabolic diseases. Diabetes Metab 2009;35(4):262–72.
[22] Galassi A, Reynolds K, He J. Metabolic syndrome and risk of cardiovascular disease: a meta-analysis. Am J Med 2006;119(10):812–9.
[23] Koren O, Spor A, Felin J, Fak F, Stombaugh J, Tremaroli V, et al. Human oral, gut, and plaque microbiota in patients with atherosclerosis. Proc Natl Acad Sci USA 2011;108(Suppl. 1):4592–8.
[24] Greenland P. ACCF/AHA guideline for assessment of cardiovascular risk in asymptomatic adults: a report of the American College of Cardiology Foundation/American Heart Association Task Force on Practice Guidelines. J Am Coll Cardiol 2010;56: e50–103.
[25] Ellis CL, Bokulich NA, Kalanetra KM, Mirmiran M, Elumalai J, Haapanen L, et al. Probiotic administration in congenital heart disease: a pilot study. J Perinatol: Off J Calif Perinat Assoc 2013;33(9):691–7.
[26] Berni Canani R, Costanzo MD, Leone L, Pedata M, Meli R, Calignano A. Potential beneficial effects of butyrate in intestinal and extraintestinal diseases. World J Gastroenterol 2011;17(12):1519–28.
[27] Trompette A, Gollwitzer ES. Gut microbiota metabolism of dietary fiber influences allergic airway disease and hematopoiesis. Nat. Med 2014;20:159–66.
[28] Louis P, Flint HJ. Diversity, metabolism and microbial ecology of butyrate-producing bacteria from the human large intestine. FEMS Microbiol Lett 2009;294: 1–8.
[29] Karlsson FH, Fåk F, Nookaew I, Tremaroli V, Fagerberg B, Petranovic D, et al. Symptomatic atherosclerosis is associated with an altered gut metagenome. Nat Commun 2012;3:1245.
[30] Karlsson FH, et al. Gut metagenome in European women with normal, impaired and diabetic glucose control. Nature 2013;498(7452):99–103.
[31] Qin J, Li Y, Cai Z, Li S, Zhu J, Zhang F, et al. A metagenome-wide association study of gut microbiota in type 2 diabetes. Nature 2012;490(7418):55–60.

[32] Furet JP, Kong LC, Tap J, Poitou C, Basdevant A, Bouillot JL, et al. Differential adaptation of human gut microbiota to bariatric surgery-induced weight loss: links with metabolic and low-grade inflammation markers. Diabetes 2010;59(12):3049–57.
[33] Zhang X, Shen D, Fang Z, Jie Z, Qiu X, Zhang C, et al. Human gut microbiota changes reveal the progression of glucose intolerance. PLoS One 2013;8(8):e71108.
[34] Remely M, Aumueller E, Merold C, Dworzak S, Hippe B, Zanner J, et al. Effects of short chain fatty acid producing bacteria on epigenetic regulation of FFAR3 in type 2 diabetes and obesity. Gene 2014;537(1):85–92.
[35] Wachtershauser A, Stein J. Rationale for the luminal provision of butyrate in intestinal diseases. Eur J Nutr 2000;39(4):164–71.
[36] Halestrap AP. The monocarboxylate transporter family—structure and functional characterization. IUBMB Life 2012;64(1):1–9.
[37] Kim MH, Kang SG, Park JH, Yanagisawa M, Kim CH. Short-chain fatty acids activate GPR41 and GPR43 on intestinal epithelial cells to promote inflammatory responses in mice. Gastroenterology 2013;145(2):396–406.
[38] Maslowski KM, Vieira AT, Ng A, Kranich J, Sierro F, Yu D, et al. Regulation of inflammatory responses by gut microbiota and chemoattractant receptor GPR43. Nature 2009;461(7268):1282–6.
[39] Brown AJ, Goldsworthy SM, Barnes AA, Eilert MM, Tcheang L, Daniels D, et al. The Orphan G protein-coupled receptors GPR41 and GPR43 are activated bypropionate and other short chain carboxylic acids. J Biol Chem 2003;278(13):11312–9.
[40] Le Poul E, Loison C, Struyf S, Springael JY, Lannoy V, Decobecq ME, et al. Functional characterization of human receptors for short chain fatty acids and their role in polymorphonuclear cell activation. J Biol Chem 2003;278(28):25481–9.
[41] Li L, Hua Y, Ren J. Short-chain fatty acid propionate alleviates Akt2 knockout-induced myocardial contractile dysfunction. Exp Diabetes Res 2012;2012:851717.
[42] Brown AJ, Goldsworthy SM, Barnes AA, Eilert MM, Tcheang L, Daniels D, et al. The orphan G protein-coupled receptors GPR41 and GPR43 are activated by propionate and other short chain carboxylic acids. J Biol Chem 2003;278:11312–9.
[43] Lannoy V, Decobecq ME, et al. Functional characterization of human receptors for short chain fatty acids and their role in polymorphonuclear cell activation. J Biol Chem 2003;278(28):25481–9.
[44] Parodi PW. Cooperative action of bioactive components in milk fat with PPARs may explain its anti-diabetogenic properties. Med Hypotheses 2016;89:1–7.
[45] Russo R, De Caro C, Avagliano C, Cristiano C, La Rana G, Mattace Raso G, et al. Sodium butyrate and its synthetic amide derivative modulate nociceptive behaviors in mice. Pharmacol Res 2016;103:279–91.
[46] Phipps RP, Blumberg N. Statin islands and PPAR ligands in platelets. Arterioscler Thromb Vasc Biol 2009;29(5):620–1.
[47] Lee JJ, Jin YR, Yu JY, Munkhtsetseg T, Park ES, Lim Y, et al. Antithrombotic and antiplatelet activities of fenofibrate, a lipid-lowering drug. Atherosclerosis 2009;206(2):375–82.
[48] Ali FY, Armstrong PC, Dhanji AR, Tucker AT, Paul-Clark MJ, Mitchell JA, et al. Antiplatelet actions of statins and fibrates are mediated by PPARs. Arterioscler Thromb Vasc Biol 2009;29(5):706–11.
[49] Liu ZM, Hu M, Chan P, Tomlinson B. Early investigational drugs targeting PPAR-α for the treatment of metabolic disease. Expert Opin Investig Drugs 2015;24(5):611–21.
[50] Mujumdar VS, Tummalapalli CM, Aru GM, Tyagi SC. Mechanism of constrictive vascular remodeling by homocysteine: role of PPAR. Am J Physiol Cell Physiol 2002;282(5):C1009–15.

[51] Diep QN, Benkirane K, Amiri F, Cohn JS, Endemann D, Schiffrin EL. PPAR alpha activator fenofibrate inhibits myocardial inflammation and fibrosis in angiotensin II-infused rats. J Mol Cell Cardiol 2004;36(2):295–304.
[52] Berni Canani R, Costanzo MD, Leone L, Bedogni G, Brambilla P, Cianfarani S, et al. Epigenetic mechanisms elicited by nutrition in early life. Nutr Res Rev 2011;24(2):198–205.
[53] Li CJ, Li RW, Baldwin RL, Elsasser TH. Butyrate: A Dietary Inhibitor of Histone Deacetylases and an Epigenetic Regulator. In: Li Cong-Jun, editor. Butyrate food source, functions and health benefits. New York: ConNova Science Publishers, Inc.; 2014. p. 233–8.
[54] Xiong Y, Dowdy SC, Podratz KC, Jin F, Attewell JR, Eberhardt NL, et al. Histone deacetylase inhibitors decrease DNA methyltransferase-3B messenger RNA stability and down-regulate de novo DNA methyltransferase activity in human endometrial cells. Cancer Res 2005;65(7):2684–9.
[55] Li RW, Li CJ. Enhancing Butyrate Biosynthesis in the Gut for Health Benefits. In: Li Cong-Jun, editor. Butyrate food source, functions and health benefits. ConNova Science PublishersPL New York; 2014. p. 1–23.
[56] Antos CL, McKinsey TA, Dreitz M, Hollingsworth LM, Zhang CL, Schreiber K, et al. Dose-dependent blockade to cardiomyocyte hypertrophy by histone deacetylase inhibitors. J Biol Chem 2003;278(31):28930–7.
[57] Gallo P, Latronico MV, Gallo P, Grimaldi S, Borgia F, Todaro M, et al. Inhibition of class I histone deacetylase with an apicidin derivative prevents cardiac hypertrophy and failure. Cardiovasc Res 2008;80(3):416–24.
[58] Granger A, Abdullah I, Huebner F, Stout A, Wang T, Huebner T, et al. Histone deacetylase inhibition reduces myocardial ischemia-reperfusion injury in mice. FASEB J 2008;22(10):3549–60.
[59] Kong Y, Tannous P, Lu G, Berenji K, Rothermel BA, Olson EN, et al. Suppression of class I and II histone deacetylases blunts pressure-overload cardiac hypertrophy. Circulation 2006;113(22):2579–88.
[60] Zhao TC, Cheng G, Zhang LX, Tseng YT, Padbury JF. Inhibition of histone deacetylases triggers pharmacologic preconditioning effects against myocardial ischemic injury. Cardiovasc Res 2007;76(3):473–81.
[61] Cavasin MA, Demos-Davies K, Horn TR, Walker LA, Lemon DD, Birdsey N, et al. Selective class I histone deacetylase inhibition suppresses hypoxia-induced cardiopulmonary remodeling through an antiproliferative mechanism. Circ Res 2012;110:739–48.
[62] Cavasin MA, Stenmark KR, McKinsey TA. Emerging roles for histone deacetylases in pulmonary hypertension and right ventricular remodeling (2013 Grover Conference series). Pulm Circ 2015;5(1):63–72.
[63] Ho KJ, Xiong L, Hubert NJ, Nadimpalli A, Wun K, Chang EB, et al. Vancomycin treatment and butyrate supplementation modulate gut microbe composition and severity of neointimal hyperplasia after arterial injury. Physiol Rep 2015;3(12).
[64] Yideng J, Jianzhong Z, Ying H, et al. Homocysteine-mediated expression of SAHH, DNMTs, MBD2, and DNA hypomethylation potential pathogenic mechanism in VSMCs. DNA Cell Biol 2007;26:603–11.
[65] Jiang Y, Sun T, Xiong J, et al. Hyperhomocysteinemia-mediated DNA hypomethylation and its potential epigenetic role in rats. Acta Biochim Biophys Sin 2007;39:657–67.
[66] Yideng J, Zhihong L, Jiantuan X, et al. Homocysteine-mediated PPARalpha,gamma DNA methylation and its potential pathogenic mechanism in monocytes. DNA Cell Biol 2008;27:143–50.
[67] Baccarelli A, Bollati V. Epigenetics and environmental chemicals. Curr Opin Pediatr 2009;21:243–51.

[68] Hortobagyi GN. Anthracyclines in the treatment of cancer. An overview. Drugs 1997;54(Suppl. 4):1–7.
[69] Young RC, Ozols RF, Myers CE. The anthracycline antineoplastic drugs. N Engl J Med 1981;305(3):139–53.
[70] Rephaeli A, Waks-Yona S, Nudelman A, Tarasenko I, Tarasenko N, Phillips DR, et al. Anticancer prodrugs of butyric acid and formaldehyde protect against doxorubicin-induced cardiotoxicity. Br J Cancer 2007;96(11):1667–74.
[71] Tarasenko AS, Sivko RV, Krisanova NV, Himmelreich NH, Borisova TA. Cholesterol depletion from the plasma membrane impairs proton and glutamate storage in synaptic vesicles of nerve terminals. J Mol Neurosci 2010;41(3):358–67.
[72] Tarasenko N, Kessler-Icekson G, Boer P, Inbal A, Schlesinger H, Phillips DR, et al. The histone deacetylase inhibitor butyroyloxymethyl diethylphosphate (AN-7) protects normal cells against toxicity of anticancer agents while augmenting their anticancer activity. Invest New Drugs 2012;30(1):130–43.

CHAPTER 5

The Relationship Between Probiotics and Dietary Fiber Consumption and Cardiovascular Health

Puttur D. Prasad, Ashish Gurav, Huabin Zhu, Pamela M. Martin, Matam Vijay-Kumar, Nagendra Singh

1 INTRODUCTION

According to Centers for Disease Control and Prevention (CDC), approximately 610,000 deaths per year (~25% of all deaths) in the United States of America are attributed to cardiovascular diseases. Hypertension, increased body weight, obesity, and increased cholesterol (hypercholesterolemia) as well as glucose levels (hyperglycemia or diabetes) are major risk factors for cardiovascular diseases. Homeostasis of glucose and cholesterol are significantly regulated by dietary intakes. Diets rich in simple sugars, cholesterol, and saturated fatty acids increase levels of glucose, cholesterol, and triglyceride in blood and thus worsen the cardiovascular diseases. On the other hand, several lines of evidence indicate that consumption of probiotics and prebiotics decrease risk for cardiovascular diseases specifically by improving blood cholesterol and glucose levels. Probiotics are live microorganisms that, upon intake, stimulate health benefits to the host. A complex consortium of bacteria and other microbes, which are collectively called gut microbiota, resides in a healthy gut. Prebiotics are indigestible food components that upon intake stimulate growth, composition, and activity of beneficial gut microbiota, which overall improves the health of the host. Dietary fibers are prebiotics and favor selective growth of specific constituents of gut microbiota that promote cardiovascular health [1]. Similarly, intake of specific members of gut microbiota mostly belonging to genus *Lactobacillus* or *Bifidobacterium* also decrease risk factors associated with cardiovascular diseases. Moreover, consumption of prebiotics and probiotics which are together

Dietary Fiber for the Prevention of Cardiovascular Disease
http://dx.doi.org/10.1016/B978-0-12-805130-6.00005-7

called synbiotics, also improves blood glucose level and lipid profile. In this chapter we will summarize the recent progress in our understanding of dietary fiber and probiotics in improving cardiovascular health and underlying mechanisms.

2 DIETARY FIBERS

E.H. Hipsley, first conceived the term "dietary fiber" in 1953 [2]. Mammals are unable to digest dietary fiber in the small intestine due to absence of specific enzymes. However, when the undigested dietary fiber reach the large intestine where a large number of gut microbiota are equipped with a spectrum of carbohydrases reside, dietary fibers get digested into numerous metabolites in the lumen. Short-chain fatty acids (SCFAs) are the major metabolites of dietary fiber degradation. Acetate, propionate, and butyrate are the most abundant SCFAs and are generated at a collective concentration of approximately 100 mM in the colonic lumen. The amounts of acetate, propionate, and butyrate in colonic lumen is approximately 60:20:20, respectively [3].

Dietary fibers are a large group of heterogeneous polysaccharides that are largely derived from plant cell walls that is poorly digested by mammals. Depending on length of the monomer chain, dietary fibers are classified into short-chain and long-chain carbohydrates [4]. Oligosaccharides made of fewer monomers of simple sugars, such as glucose, fructose, or galactose are called short-chain dietary fibers. They are highly soluble and easily fermentable. Some long-chain carbohydrates, such as pectin (polygalactouronate), guar gum (galactomannan), and inulin (polyfructose) are also soluble and highly fermentable [4]. On the other hand, other long-chain dietary fibers, such as cellulose and methylcellulose are insoluble and nonfermentable. A single food source, such as legumes, wheat, onion, artichoke, or oats contains a mixture of short- to long-chain dietary fibers, with varying degrees of fermentability. Soluble dietary fibers are fermented by gut microbiota. The insoluble part of dietary fiber is neither digested nor fermented, and thus mostly serve as fecal bulk having a laxative effect by facilitating proper bowel movement.

Fermentable dietary fibers have been used in food supplements as "prebiotics" to positively influence the gut ecology and homeostasis by promoting composition and metabolic activity of gut microbiota, resulting in multiple beneficial effects, such as cardiovascular health on host. Fermentation of dietary fiber into different SCFAs is a stepwise process and available evidence indicates that distinct groups of gut bacteria are involved at each step. For example, in vitro studies demonstrate that fermentation of dietary fiber by *Bifidobacterium* results in production of acetate, fructose, and lactate, but little butyrate [5]. Butyrate-producing gut bacteria, such as *Roseburia intestinalis* and *Anaerostipes caccae* are poor fermenters of dietary fiber [5]. However, in mixed cultures containing dietary fiber, *Bifidobacterium* and *R. intestinalis* or *A. caccae*, both acetate and butyrate are produced [5]. Mechanistic studies reveal that addition of acetate leads to fermentation of dietary fiber by *R. intestinalis* and production of butyrate. This phenomenon is termed as "cross-feeding," where metabolites produced by one group of organisms are used by another group for their growth. An evidence of cross feeding in vivo is demonstrated by a recent human study which shows that dietary fiber consumption enhances cooccurrence between potential butyrate producers and gut bacteria that directly ferment complex carbohydrates [6].

3 DIETARY FIBER AND CARDIOVASCULAR HEALTH

Epidemiological studies suggest that intake of 14 g of dietary fiber per 1000 kcal or 25 g for adult women and 28 g for adult men is significantly associated with reduced risk of cardiovascular diseases [7]. Fermentable dietary fibers, such as inulin, pectin, psyllium, guar gum, and oligofructose have generated lots of interest due to their ability to improve blood lipid profile. Many studies have evaluated effectiveness of dietary fibers in reducing cholesterol levels in hypercholesterolemic/hyperlipidemic individuals. A double-blinded, randomized clinical study examined the effect of inulin in 12 dyslipidemic obese subjects between 19 and 32 years of age. Subjects received 7 g/day of inulin or placebo in the morning. After 4 weeks, a significant reduction in total cholesterol, low-density lipoprotein (LDL) cholesterol, very low-density lipoprotein (VLDL), and triglyceride levels in blood was observed [8]. Similarly, in another double-blinded placebo-controlled study, obese and moderately dyslipidemic premenopausal women were given two doses of yacon syrup containing 0.29 g and 0.14 g fructooligosaccharides (FOS)/kg/day for 4 months. Yacon syrup significantly decreased body weight, waist circumference, body mass index (BMI), fasting serum insulin, and LDL cholesterol, whereas fasting glucose and triglycerides were not affected [9]. In a separate randomized double-blinded study, 12 hypercholesterolemic men consumed 1 pint of vanilla ice cream with or without 20 g of inulin from chicory root [degree of polymerization (DP) between 2 and 60]. In the group consuming inulin, serum triglycerides decreased by 40 mg/dL ($P < 0.05$) [10]. The beneficial effect of dietary fiber was also observed in type II diabetic individuals. Consumption of 8 g of FOS significantly decreased average fasting blood glucose, total cholesterol, and LDL-cholesterol levels by 15, 19, 17 mg/dL, respectively. However, serum HDL cholesterol, triglyceride, or free fatty acid (FFA) levels were not changed by FOS in this study [11]. In contrast to these encouraging results, 15 and 20 g of daily intake of FOS for 3 and 4 weeks, respectively did not significantly alter the blood glucose, lipid profile, or insulin resistance in type II diabetes patients [12,13]. Collectively, these findings suggest that more detailed studies are needed to understand the effectiveness of dietary fibers in reducing cardiovascular risk.

There have been studies conducted to test effectiveness of dietary fiber in improving blood lipid profiles in healthy individuals. In a randomized, double-blinded, placebo-controlled and parallel design, healthy young individuals consumed 50 g of a rice-based ready-to-eat cereal daily with or without 18% inulin for breakfast. After 12 weeks, the group consuming inulin showed significant decrease in plasma total cholesterol and triglycerides by 7.9 ± 5.4% ($P < 0.05$) and 21.2 ± 7.8% ($P < 0.005$), respectively [14]. Another randomized and placebo-controlled crossover study examined the effect of inulin in changing the lipid profile in healthy men. Individuals were given moderately high-carbohydrate, low-fat diet (55% of total energy) plus an oral placebo or 10 g of high-performance inulin daily. This study found that, compared to placebo, inulin significantly reduced plasma triglycerides and hepatic lipogenesis ($P < 0.05$) in 3 weeks [15]. However, cholesterol synthesis and plasma cholesterol concentrations were not significantly different between placebo and inulin group. Similarly, in another study involving 22 healthy individuals, inulin significantly improved HDL cholesterol, total cholesterol/HDL-cholesterol ratio, triglycerides, and lipoproteins [16]. In contrast, other studies failed to find the effectiveness of inulin in changing lipid profiles in healthy individuals [17–20].

In contrast with human studies, several animal studies have consistently showed beneficial effect of dietary fiber in changing blood lipid profile, improving blood glucose tolerance, and insulin resistance [21,22]. A critical difference between human and animal studies is the amount of dietary fiber consumed per day. Most of the human studies have used 5–20 g of daily intake of dietary fiber per individual. Considering ~70 kg body weight for an individual, this equals to approximately 0.07–0.28 g of dietary fiber per kg of body weight. Studies conducted in rats use 10%–20% of inulin in animal chow diet. A rat weighs around 300 g and eats approximately 10–30 g of food per day. Therefore, a typical rat study uses ~3.3–10 g of dietary fiber per kg of body weight [23]. It is difficult to increase daily intake of dietary fiber in humans to this level because raising the amount of dietary fiber is associated with adverse side effects, such as gastrointestinal cramps, bloating, and diarrhea [19]. The other variables, such as diverse genetic background, life style, medical history, background diet may also affect the effectiveness of dietary fiber in changing lipid profiles in humans. Several human studies have used different vehicles for delivery of dietary fiber, such as vanilla ice cream, syrup, margarine, breakfast cereal, orange juice, and pasta. Moreover, use of different types of dietary fiber may also influence the outcomes observed in human studies.

4 PROBIOTICS AND CARDIOVASCULAR HEALTH

The word probiotics means "for life." Most of the probiotics are part of normal gut microbiota under steady state conditions. At present, commonly used probiotics are *Lactobacillus casei*, *Lactobacillus reuteri*, *Lactobacillus acidophilus*, *Lactobacillus rhamnosus*, *Lactobacillus plantarum*, *Bifidobacterium infantis*, *Bifidobacterium longum*, *Enterococcus faecium* SF68, and *Saccharomyces boulardii*. Probiotics are taken as single organism or in combination. Fermented milk or yogurt contains different species of *Lactobacillus*, and is a good source of probiotics. In 1970s, it was observed that yogurt consumption in large amounts reduces blood cholesterol level in dairying nomadic population [24,25]. In addition to fermented and unfermented milk components, yogurt also contains bacteria. Identity and composition of these bacteria varies between different manufactures. Later studies have indicated that the ability of yogurt to reduce cholesterol level depends on specific bacteria present in the yogurt [26]. Yogurt containing *L. acidophilus* L1, a bacteria derived from human gut decreased blood cholesterol in hypercholesterolemic individuals, whereas yogurt containing *L. acidophilus* ATC43121, a swine-derived gut bacteria failed to do so [26]. These findings prompted to test whether bacterial strains present in yogurt could lower blood cholesterol independent of yogurt. Indeed, consumption of lyophilized and microencapsulated *L. reuteri* NCIMB 30242 at a dose of 4×10^9 cfu/day for 9 weeks resulted in an approximate decrease in total cholesterol by 9.14%, LDL cholesterol by 11.54%, non-HDL cholesterol by 11.3%, and apoB-100 by 8.41% compared to placebo in hypercholesterolemic women and men [27]. In addition, *Lactobacillus gasseri* SBT2055

(LGSP) reduced abdominal viscera and subcutaneous fat in overweight individuals [28]. Perinatal as well as postnatal consumption of *L. rhamnosus* GG significantly decreased weight gain in children [29]. In a relatively small number of clinical trials, probiotics failed to improve the lipid profile in hypercholesterolemic individuals. For example, *Lactobacillus fermentum* at a dose of 2 capsules (2×10^9 cfu per capsule) 2 times per day for 10 weeks did not change blood cholesterol in hypercholesterolemic subjects [30]. In contrast to high success in reducing cholesterol levels in hypercholesterolemic individuals, probiotics have shown limited success in improving blood cholesterol in normal individuals [31–34].

Probiotics belonging to genus other than *Lactobacillus* have also been tested to evaluate their effects on blood cholesterol levels. Human subjects were given *E. faecium* M-74 at a dose of 2×0^9 cfu/day with organically bound selenium (50 mg/dose) or placebo. After 60 weeks, the group receiving *E. faecium* M-74 showed a significant reduction in LDL cholesterol; however, no changes in HDL cholesterol or triglycerides were observed [35]. Few studies have examined the effects of a combination of two or more probiotics on cardiovascular health. A meta analysis using data from six studies showed that yogurt containing *E. faecium* and two strains of *Streptococcus thermophilus* reduced blood LDL cholesterol in 4–8 weeks [36]. In a randomized double-blinded clinical trial, yogurt containing *L. acidophilus* La5 and *Bifidobacterium lactis* Bb12 (1×10^6 cfu/day) for 6 weeks lead to significant reduction in LDL cholesterol and total cholesterol by 7.55 and 4.5%, respectively, over placebo group in type II diabetic individuals [37]. VSL#3 is a combination of *B. longum*, *B. infantis*, *Bifidobacterium breve*, *L. acidophilus*, *L. casei*, *Lactobacillus delbrueckii* ssp. *bulgaricus*, *L. plantarum,* and *Streptococcus salivarius* ssp. *thermophilus*. VSL#3 administration significantly decreased triglycerides, C-reactive protein and increased HDL cholesterol without affecting levels of LDL cholesterol and total cholesterol in critically ill patients [38]. Similarly, in a double-blinded randomized clinical trial involving obese nonalcoholic fatty liver disease (NAFLD) children, VSL#3 administration decreased BMI and improved NAFLD [39]. Moreover, in healthy men consuming high-fat diet, VSL#3 reduced BMI, suggesting a preventive role of this probiotic combination in obesity.

Probiotics have also been effective in decreasing body weight and improving cardiovascular health in animals. Conjugated linoleic acid (CLA) -producing probiotics, such as *L. rhamnosus* PL60 decreased body weight and adipose mass without affecting the food intake in obese mice [40].

Similarly, *L. plantarum* PL62 reduced high-fat-diet-induced weight gain and blood glucose in obese mice [41]. Certain probiotics, such as LGSP and *L. plantarum* strain No. 14 have been shown to reduce the size of adipocytes in animal studies [42,43].

5 SYNBIOTICS

Since both prebiotics and probiotics reduce the risk factors associated with cardiovascular diseases, several studies have examined the effects of a combination of prebiotics and probiotics, which are together called synbiotics, in reducing cardiovascular disease-associated risk factors. In a randomized placebo-controlled double-blinded two-way crossover trial, intake of milk fermented with yogurt starter *L. acidophilus*, which contained 2.5% FOS, 0.5% vegetable oil, and 0.5% milk fat 3 times/day, significantly lowered total cholesterol and LDL cholesterol by 4.4 and 5.4%, respectively than milk fermented with yogurt strains and containing 1% milk fat [44]. Similarly, in another randomized, double-blinded, placebo-controlled, and parallel-designed study involving hypercholesterolemic men and women, consumption of four capsules containing *L. gasseri* CHO-220 (10^9 cfu/day) and inulin (0.8 g) per day for 12 weeks leads to significant reduction in plasma total cholesterol and LDL cholesterol by 7.84 and 9.27%, respectively than placebo. Additionally, in the synbiotic group, a decrease of triglyceride was observed in all three lipoproteins, VLDL, LDL, and HDL [45]. In contrast, other studies using *L. acidophilus* and *B. longum* and 10–15 mg FOS or yogurt enriched with *L. acidophilus* 145, *B. longum* 913, and 1% oligofructose did not change the plasma levels of LDL and total cholesterol [46,47]. However, increase in HDL cholesterol was observed in the latter study. Collectively, these findings show the effects of synbiotics in improving the lipid profile in the blood may be strain-specific and additional studies are needed to determine their efficacy in a larger population.

6 MECHANISMS OF ACTION

6.1 Mechanism of Lipid-Lowering Effects of Dietary Fiber

6.1.1 Composition of Gut Microbiota

Dietary fiber acts as a nutrient for gut microbiota and thus promote growth of beneficial gut bacteria, such as *Bifidobacterium,* which are known to reduce risk of cardiovascular diseases [48]. However, molecular mechanism(s) underlying selective promotion of *Bifidobacterium* by dietary fiber remain poorly understood.

6.1.2 Absorption and Excretion of Lipids

Soluble dietary fibers decrease uptake of lipids and cholesterol in intestine (Fig. 5.1) [49–51]. It is believed that viscosity produced by soluble fiber play a key role in reducing cholesterol absorption in intestine. Dietary fiber binds to cholesterol directly, reduce diffusion of cholesterol toward apical surface of intestinal mucosal cells, and interfere with emulsification of cholesterol in the intestinal lumen [52,53]. In addition, dietary fibers decrease the enterohepatic pool of bile acids by enhancing the excretion of bile acids [54]. In human studies, native amylo-maize-resistant starch promotes the fecal excretion of bile acids [55]. To compensate the loss of cholesterol, liver consumes cholesterol from systemic circulation and produces more bile acid, resulting in decreased blood cholesterol levels [56]. Consumption of guar gum increases expression of hepatic HMG-CoA reductase, the rate-limiting enzyme in cholesterol biosynthesis. However, guar gum-induced fecal loss of cholesterol overwhelms the cholesterol biosynthesis, resulting in reduced levels of cholesterol [57].

6.1.3 Changing Metabolism

Inulin intake reduces lipogenesis in the liver and fasting triglyceride levels in healthy individuals without affecting blood cholesterol levels (Fig. 5.1) [15]. The decrease in lipogenesis is due to dietary fiber-mediated suppression of lipogenic enzymes, such as acetyl-CoA carboxylase (ACC), fatty acid synthase (FAS), malic enzyme, ATP citrate lyase, and glucose-6-phosphate dehydrogenase in liver [58]. However, de novo lipogenesis is greatly influenced by background diet. Oligofructose induces secretion of glucose-dependent insulinotrophic polypeptide (GIP) in rats [22]. GIP is known to enhance activity of lipoprotein lipase (LPL), the key enzyme involved in the clearance of triglyceride-enriched lipoproteins following lipid intake, which will lead to a reduction in plasma triglyceride levels [59]. In animal studies, both β-glucagon and inulin decreased body weight. Interestingly,

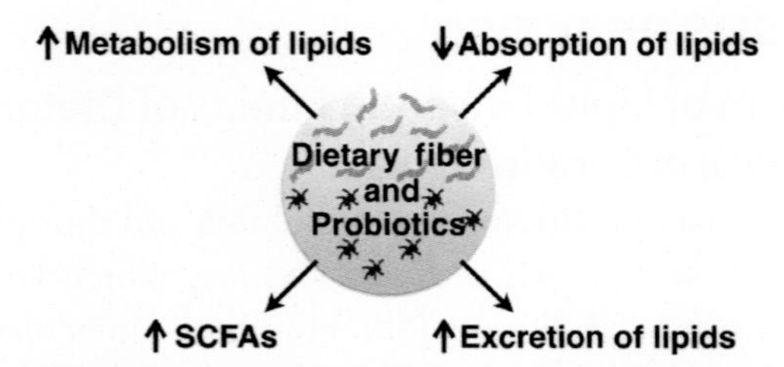

Figure 5.1 ***Possible mechanisms underlying the beneficial effects of probiotics and dietary fiber in cardiovascular health.***

inulin decreased body adiposity without any effect on food intake, whereas, β-glucagon decreased food intake [60]. These findings suggest that different dietary fibers decrease cardiovascular risks by stimulating different mechanisms.

6.1.4 Production of SCFAs

Several pieces of evidence demonstrate a critical role of SCFAs, which are major metabolites of dietary fibers, in regulating cardiovascular risk factors (Fig. 5.1). Butyrate is metabolized by colonic epithelium for their energy needs and converted into ketone bodies [61]. Acetate and propionate are transported via portal circulation into liver. Acetate is converted into acetyl-CoA and acts as precursor for lipogenesis [62,63]. In isolated rat hepatocytes, propionate inhibits incorporation of acetate into fatty acids and cholesterol [62]. In accordance with this, in a human study, rectal infusion of acetate and propionate inhibited incorporation of acetate into plasma lipids [64]. Toll-like receptor 5 (TLR5)-deficient mice develop microbiota-dependent metabolic syndrome. SCFAs increase expression of stearoyl-CoA desaturase 1 (SCD1) in *Tlr5*$^{-/-}$ mice leading to enhanced lipogenesis and worsening of metabolic syndrome in these mice [65].

G-protein coupled cell surface receptors GPR41, GPR43, and GPR109A bind to SCFAs and have been actively explored as a link between dietary fiber and reduction in cardiovascular risk. GPR41 and GPR43 bind to all SCFAs, whereas GPR109A binds to butyrate only. GPR41 and GPR43 are expressed by enteroendocrine cells of intestine, and induce secretion of anorexic hormones PYY and glucagon-like peptide 1 (GLP-1) [66]. Release of PYY into blood decreases appetite and thus inhibits obesity. GLP-1 induces production of insulin by pancreas leading to increased metabolism of blood glucose and development of glucose tolerance. It was noted that *Gpr41*$^{-/-}$ mice possess less body fat than WT counterparts [67]. Another study found that both *Gpr41*$^{-/-}$ and WT mice contain similar amounts of fat [68]. Opposing role of GPR43 in regulation of obesity has also been observed. One study demonstrated that on a high-fat diet, *Gpr43*$^{-/-}$ mice were leaner than WT controls [69]. However, on a normal diet *Gpr43*$^{-/-}$ mice gained more weight than WT littermates [70]. In a separate study, all three SCFAs, acetate, propionate, and butyrate protected mice against high-fat diet-induced obesity and insulin resistance [71]. Notably, effect of butyrate and propionate was more pronounced than acetate. In addition, butyrate and propionate also induced secretion of glucose-dependent insulinotrophic peptide, amylin, and insulin. Surprisingly, these effects were independent of Gpr41. Role of Gpr43 was not evaluated in this study. GPR109A is expressed at highest level in adipocytes and innate immune cells, such as macrophages, granulocytes, and dendritic cells. Activation of GPR109A and GPR43 on adipocytes inhibits activation of hormone sensitive lipase, an enzyme that catalyzes degradation of triglycerides in adipose tissue into FFAs and glycerol, both of which are subsequently released into blood. Therefore, GPR109A and GPR43 signaling have the ability to decrease plasma FFA levels. Niacin-mediated lowering of FFAs and triglycerides is at least partially dependent on GPR109A [72,73]. Activation of GPR109A on surface of macrophages plays a critical role in induction of cholesterol transporter ABCG1 leading to increased cholesterol efflux [74]. However, role of GPR109A in dietary fiber-mediated improvement of blood cholesterol and triglyceride levels has not being investigated yet. Collectively, these findings indicate that a more careful evaluation of GPR41, GPR43, and GPR109A is needed to establish their role in obesity and metabolic diseases.

6.2 Mechanism of Lipid-Lowering Effects of Probiotics

A number of different mechanisms that link probiotics to a decrease in cardiovascular risk factors have been proposed (Fig. 5.1). Studies conducted several decades ago observed that conventional rats exhibit lower accumulation of intravenously injected cholesterol-26-(^{14})C in blood and liver than germ-free rats [75]. The reduction in cholesterol was due to its accelerated excretion in urine and feces. Additional studies also showed that germ-free animals exhibit ~25% higher absorption of dietary cholesterol in intestine than conventional rats [76]. Accordingly, germ-free rats fed with moderate amounts of cholesterol showed 2–3 times higher levels of cholesterol than conventional rats. These earlier observations pioneered research investigating the role of intestinal bacteria and probiotics in regulating cholesterol metabolism in host.

6.2.1 Reduced Absorption of Lipids in Intestine

Niemann-Pick C1-like 1 (NPC1L1) protein is a sterol transporter localized at the apical membrane of intestinal epithelium and thus mediates intestinal cholesterol absorption. *L. acidophilus* ATCC 4356 decreases expression of NPC1L1 by a mechanism involving liver X receptors (LXR) leading to reduced cholesterol absorption in intestine [77]. LGSP binds to intestinal lipids and cholesterol and thus inhibits intestinal absorption of lipids [78].

This binding is strain and growth phase specific. Heat-killed *Lactobacillus lactis* subsp. *lactis* biovar *diacetylactis* N7 binds to cholesterol albeit at lower efficiency than live bacteria [79]. In addition, active metabolism of cholesterol by probiotics may also contribute to their lipid-lowering effects on host. Growing *L. lactis* subsp. *lactis* biovar *diacetylactis* N7 also incorporates cholesterol in its membrane, which could explain cholesterol-lowering ability of this probiotic [79].

6.2.2 Excretion of Lipids

LGSP promotes excretion of neutral sterols, bile acids, and inhibits lymphatic absorption of triglycerides, phospholipids, and cholesterol leading to decreased size of adipocytes and adipose mass [80]. Certain probiotics, such as *L. acidophilus* ATCC 314, *L. acidophilus* FTCC 0291, *L. bulgaricus* FTCC 0411, *L. bulgaricus* FTDC 1311, and *L. casei* ATCC 393 express intracellular as well as extracellular cholesterol reductase, an enzyme that converts cholesterol to coprostanol [81]. Coprostanol are excreted by intestine. Several probiotics, such as *L. reuteri* NCIMB 30242 express bile salt hydrolase, an enzyme that deconjugates bile acids [27]. Bile acids are made of cholesterol. Deconjugated bile acids are less soluble and are excreted in feces. Moreover, to compensate the loss of excreted bile acids, the liver metabolizes serum cholesterol and converts it into bile acids, leading to reduction of serum cholesterol.

6.2.3 Changing Metabolism in Host

Uncoupling protein-2 (UCP-2) plays a key role in energy expenditure. CLA produced by probiotics plays an important role in enhancing metabolism of host. Accordingly, CLA as well as CLA-producing probiotics, such as *L. rhamnosus* PL60 reduce adipose mass and induce expression of UCP-2 in adipose tissue of multiple mouse strains, such as C57BL/6, Ob/Ob, and ICR [82,83]. Insoluble fraction of Kefir, a traditional fermented milk, suppressed expression of adipogenic transcription factors: CCAAT/enhancer-binding protein alpha (C/EBPα), sterol regulatory element-binding protein-1 (SREBP1), and peroxisome proliferator-activated receptor gamma (PPARγ) and thus inhibits differentiation of 3T3-L1, a preadipocyte, into adipocyte in vitro [84]. A combination of *L. rhamnosus GG* and *Lactobacillus sakei* NR28 reduced expression of spectrum of enzymes involved in hepatic lipogenesis, such as ACC, FAS, and SCD1, which explain the ability of this probiotic to decrease weight gain [85]. LPL is a key enzyme that promotes import of triglycerides to adipose tissue for storage. *Lactobacillus paracasei* ssp. *paracasei* F19 (F19) increases levels of angiopoietin-like 4 (ANGPTL4), an inhibitor of LPL, which may explain the ability of this probiotic to decrease adipose mass and reduction of triglycerides in different fractions of lipoproteins [86]. These findings indicate that the mechanisms underlying improvement of cardiovascular health by probiotics may be strain specific. The evidence for this possibility was shown by a study which compared three different strains of *L. reuteri,* ATCC PTA 4659 (ATCC), DSM 17938 (DSM), and L6798 to reduce cardiovascular risk in atherosclerosis prone *apoE*$^{-/-}$ mice. Only ATCC strain increased levels of carnitine palmitoyltransferase 1a (CPT1a), which plays a very important role in fatty acid β-oxidation in the liver, resulting in decreased fat accumulation in the liver and reduced body weight [87]. Other strains of *L. reuteri* used in this study were ineffective in changing any parameters associated with atherosclerosis. VSL#3 increased the levels of GLP-1 [39]. Activated GLP-1 (aGLP-1), which is known to decrease blood glucose level, improves insulin sensitivity and delays gastric emptying. In addition, VSL#3 also inhibits high-fat diet-induced depletion of natural killer T (NKT) cells, resulting in improved insulin resistance and hepatic steatosis [88].

6.2.4 Changing Composition of Gut Microbiota

Probiotics are live bacteria that colonize some part of the gastrointestinal tract. This colonization leads to change in composition of gut bacterial species residing in the gut. Animal studies show that composition of gut microbiota influences and regulates metabolic syndrome, a risk factor for cardiovascular disease. A number of different human and animal studies have shown that increased numbers of Firmicutes are associated with obesity [89,90]. Administration of *L. rhamnosus* GG and *Lactobacillus sakei* NR28 decreased the ratio of Firmicutes:Bacteroidetes in small intestine and lowered epididymal fat mass and biomarkers associated with metabolic syndrome, such as ACC, FAS, and SCD1 in the liver [85]. Inoculation of a different species of *Lactobacillus*, *Lactobacillus ingluviei* increased numbers of Firmicutes and gain in body weight, metabolism, inflammation, and liver size [91]. Other studies have shown opposite role of Firmicutes in regulation of obesity [92]. Taxonomically, Firmicutes are phylum of bacteria, and thus comprise of diverse cellular, molecular and metabolic profiles. Some of the members in this phylum are butyrate producers. *Clostridium butyricum* MIYAIRI 588—a butyric acid-producing gut bacteria inhibits choline-deficient/L-amino acid-defined (CDAA)-diet-induced NAFLD in rats [93].

C. butyricum MIYAIRI 588 as well as butyrate as single agent enhanced nuclear factor erythroid 2-related factor 2 (NRF2) and decreased oxidative stress in liver demonstrating the important role of butyrate in improvement of liver and cardiovascular health [94]. Effects of other members of this phylum on cardiovascular health remains uncharacterized. Identification of specific species/strains of Firmicutes and their specific effect on cardiovascular health will be the key to understand the role of this bacteria group on our well-being (Fig. 5.1).

7 CONCLUSIONS

Both epidemiological and experimental evidences indicate a key role of probiotics and prebiotics in decreasing risk factors underlying development of cardiovascular diseases. The effects of probiotics in reducing plasma levels of total and LDL-cholesterol in hypercholesterolemic individuals are evident. However, beneficial effects are dependent on specific strains used in study. Similarly, dietary fibers are effective in reducing triglyceride levels in hypercholesterolemic and type II diabetic individuals in multiple studies. Although, these studies are encouraging, inconsistent findings have also been reported. Nonetheless, not all dietary fibers are the same and have different chemical structure and properties. Different dietary fibers have diverse effects and/or use different mechanisms to induce similar effects. Varying DP for same fiber used in different studies may yield contradictory findings. Therefore, rather than a single definition, there is a greater need to re-define dietary fibers into different groups according to their functional effects on our health. Similarly, although epidemiological studies indicate a beneficial role of dietary fiber on cardiovascular health, short duration of clinical studies may limit the observed positive effects of dietary fiber. A better effect of dietary fiber in epidemiological studies may also result from synergistic effect of a combination of dietary fibers as opposed to single type of dietary fiber used in most of the clinical studies. In addition, chemical properties of dietary fibers, such as DP [95], and solubility may change during its extraction from background diet, which may result in altered effects of dietary fiber in purified diet compared to whole food. Therefore, a detailed understanding of the chemical nature of dietary fibers, their effect in isolation versus part of background diet and molecular mechanism involved in probiotics- or prebiotics-mediated decrease in risk factors underlying development of cardiovascular diseases is warranted to delineate the effect of these food components on promotion of cardiovascular health.

REFERENCES

[1] Gibson GR, Scott KP, Rastall RA, Tuohy KM, Hotchkiss A, Dubert-Ferrandon A, et al. Dietary prebiotics: current status and new definition. Food Sci Technol Bull Funct Foods 2010;7:1–19.
[2] Hipsley EH. Dietary "fibre" and pregnancy toxaemia. Br Med J 1953;2:420–2.
[3] Ganapathy V, Thangaraju M, Prasad PD, Martin PM, Singh N. Transporters and receptors for short-chain fatty acids as the molecular link between colonic bacteria and the host. Curr Opin Pharmacol 2013;13:869–74.
[4] Eswaran S, Muir J, Chey WD. Fiber and functional gastrointestinal disorders. Am J Gastroenterol 2013;108:718–27.
[5] Falony G, Vlachou A, Verbrugghe K, De Vuyst L. Cross-feeding between *Bifidobacterium longum* BB536 and acetate-converting, butyrate-producing colon bacteria during growth on oligofructose. Appl Environ Microbiol 2006;72:7835–41.
[6] O'Keefe SJ, Li JV, Lahti L, Ou J, Carbonero F, Mohammed K, et al. Fat, fibre and cancer risk in African Americans and rural Africans. Nat Commun 2015;6:6342.
[7] Slavin JL. Position of the American Dietetic Association: health implications of dietary fiber. J Am Diet Assoc 2008;108:1716–31.
[8] Balcazar-Munoz BR, Martinez-Abundis E, Gonzalez-Ortiz M. [Effect of oral inulin administration on lipid profile and insulin sensitivity in subjects with obesity and dyslipidemia]. Rev Med Chil 2003;131:597–604.
[9] Genta S, Cabrera W, Habib N, Pons J, Carillo IM, Grau A, et al. Yacon syrup: beneficial effects on obesity and insulin resistance in humans. Clin Nutr 2009;28:182–7.
[10] Causey JL, Feirtag JM, Gallaher DD, Tungland BC, Slavin JL. Effects of dietary inulin on serum lipids, blood glucose and the gastrointestinal, environment in hypercholesterolemic men. Nutr Res 2000;20:191–201.
[11] Yamashita K, Kawai K, Itakura M. Effects of Fructo-oligosaccharides on blood-glucose and serum-lipids in diabetic subjects. Nutr Res 1984;4:961–6.
[12] Alles MS, de Roos NM, Bakx JC, van de Lisdonk E, Zock PL, Hautvast JGAJ. Consumption of fructooligosaccharides does not favorably affect blood glucose and serum lipid concentrations in patients with type 2 diabetes. Am J Clin Nutr 1999;69:64–9.
[13] Luo J, Van Yperselle M, Rizkalla SW, Rossi F, Bornet FRJ, Slama G. Chronic consumption of short-chain fructooligosaccharides does not affect basal hepatic glucose production or insulin resistance in type 2 diabetics. J Nutr 2000;130:1572–7.
[14] Brighenti F, Casiraghi MC, Canzi E, Ferrari A. Effect of consumption of a ready-to-eat breakfast cereal containing inulin on the intestinal milieu and blood lipids in healthy male volunteers. Eur J Clin Nutr 1999;53:726–33.
[15] Letexier D, Diraison F, Beylot M. Addition of inulin to a moderately high-carbohydrate diet reduces hepatic lipogenesis and plasma triacylglycerol concentrations in humans. Am J Clin Nutr 2003;77:559–64.
[16] Russo F, Chimienti G, Riezzo G, Pepe G, Petrosillo G, Chiloiro M, et al. Inulin-enriched pasta affects lipid profile and Lp(a) concentrations in Italian young healthy male volunteers. Eur J Nutr 2008;47:453–9.
[17] Kruse HP, Kleessen B, Blaut M. Effects of inulin on faecal bifidobacteria in human subjects. Br J Nutr 1999;82:375–82.
[18] Luo J, Rizkalla SW, Alamowitch C, Boussairi A, Blayo A, Barry JL, et al. Chronic consumption of short-chain fructooligosaccharides by healthy subjects decreased basal hepatic glucose production but had no effect on insulin-stimulated glucose metabolism. Am J Clin Nutr 1996;63:939–45.
[19] Pedersen A, Sandstrom B, VanAmelsvoort JMM. The effect of ingestion of inulin on blood lipids and gastrointestinal symptoms in healthy females. Br J Nutr 1997;78:215–22.

[20] van Dokkum W, Wezendonk B, Srikumar TS, van den Heuvel EG. Effect of nondigestible oligosaccharides on large-bowel functions, blood lipid concentrations and glucose absorption in young healthy male subjects. Eur J Clin Nutr 1999;53:1–7.
[21] Delzenne NM, Daubioul C, Neyrinck A, Lasa M, Taper HS. Inulin and oligofructose modulate lipid metabolism in animals: review of biochemical events and future prospects. Br J Nutr 2002;87:S255–9.
[22] Kok NN, Morgan LM, Williams CM, Roberfroid MB, Thissen JP, Delzenne NM. Insulin, glucagon-like peptide 1, glucose-dependent insulinotropic polypeptide and insulin-like growth factor I as putative mediators of the hypolipidemic effect of oligofructose in rats. J Nutr 1998;128:1099–103.
[23] Roberfroid M. Dietary fiber, inulin, and oligofructose—a review comparing their physiological-effects. Crit Rev Food Sci Nutr 1993;33:103–48.
[24] Hepner G, Fried R, Stjeor S, Fusetti L, Morin R. Hypocholesterolemic effect of yogurt and milk. Am J Clin Nutr 1979;32:19–24.
[25] Mann GV. Masai, milk and yogurt factor—alternative explanation. Atherosclerosis 1978;29:265.
[26] Anderson JW, Gilliland SE. Effect of fermented milk (yogurt) containing *Lactobacillus acidophilus* L1 on serum cholesterol in hypercholesterolemic humans. J Am Coll Nutr 1999;18:43–50.
[27] Jones ML, Martoni CJ, Prakash S. Cholesterol lowering and inhibition of sterol absorption by *Lactobacillus reuteri* NCIMB 30242: a randomized controlled trial. Eur J Clin Nutr 2012;66:1234–41.
[28] Kadooka Y, Sato M, Imaizumi K, Ogawa A, Ikuyama K, Akai Y, et al. Regulation of abdominal adiposity by probiotics (*Lactobacillus gasseri* SBT2055) in adults with obese tendencies in a randomized controlled trial. Eur J Clin Nutr 2010;64:636–43.
[29] Luoto R, Kalliomaki M, Laitinen K, Isolauri E. The impact of perinatal probiotic intervention on the development of overweight and obesity: follow-up study from birth to 10 years. Int J Obes 2010;34:1531–7.
[30] Simons LA, Amansec SG, Conway P. Effect of *Lactobacillus fermentum* on serum lipids in subjects with elevated serum cholesterol. Nutr Metab Cardiovasc Dis 2006;16:531–5.
[31] Agerbaek M, Gerdes LU, Richelsen B. Hypocholesterolaemic effect of a new fermented milk product in healthy middle-aged men. Eur J Clin Nutr 1995;49:346–52.
[32] de Roos NM, Schouten G, Katan MB. Yoghurt enriched with *Lactobacillus acidophilus* does not lower blood lipids in healthy men and women with normal to borderline high serum cholesterol levels. Eur J Clin Nutr 1999;53:277–80.
[33] Rossi EA, Vendramini RC, Carlos IZ, de Oliveira MG, de Valdez GF. Effect of a new fermented soy milk product on serum lipid levels in normocholesterolemic adult men. Arch Latinoam Nutr 2003;53:47–51.
[34] Sadrzadeh-Yeganeh H, Elmadfa I, Djazayery A, Jalali M, Heshmat R, Chamary M. The effects of probiotic and conventional yoghurt on lipid profile in women. Br J Nutr 2010;103:1778–83.
[35] Hlivak P, Odraska J, Ferencik M, Ebringer L, Jahnova E, Mikes Z. One-year application of probiotic strain *Enterococcus faecium* M-74 decreases serum cholesterol levels. Bratisl Lek Listy 2005;106:67–72.
[36] Agerholm-Larsen L, Bell ML, Grunwald GK, Astrup A. The effect of a probiotic milk product on plasma cholesterol: a meta-analysis of short-term intervention studies. Eur J Clin Nutr 2000;54:856–60.
[37] Ejtahed HS, Mohtadi-Nia J, Homayouni-Rad A, Niafar M, Asghari-Jafarabadi M, Mofid V, et al. Effect of probiotic yogurt containing *Lactobacillus acidophilus* and *Bifidobacterium lactis* on lipid profile in individuals with type 2 diabetes mellitus. J Dairy Sci 2011;94:3288–94.

[38] Sanaie S, Ebrahimi-Mameghani M, Mahmoodpoor A, Shadvar K, Golzari SE. Effect of a probiotic preparation (VSL#3) on cardiovascular risk parameters in critically-ill patients. J Cardiovasc Thorac Res 2013;5:67–70.
[39] Alisi A, Bedogni G, Baviera G, Giorgio V, Porro E, Paris C, et al. Randomised clinical trial: The beneficial effects of VSL#3 in obese children with non-alcoholic steatohepatitis. Aliment Pharmacol Ther 2014;39:1276–85.
[40] Lee HY, Park JH, Seok SH, Baek MW, Kim DJ, Lee KE, et al. Human originated bacteria, *Lactobacillus rhamnosus* PL60, produce conjugated linoleic acid and show anti-obesity effects in diet-induced obese mice. Biochim Biophys Acta 2006;1761:736–44.
[41] Lee K, Paek K, Lee HY, Park JH, Lee Y. Antiobesity effect of trans-10,*cis*-12-conjugated linoleic acid-producing *Lactobacillus plantarum* PL62 on diet-induced obese mice. J Appl Microbiol 2007;103:1140–6.
[42] Sato M, Uzu K, Yoshida T, Hamad EM, Kawakami H, Matsuyama H, et al. Effects of milk fermented by *Lactobacillus gasseri* SBT2055 on adipocyte size in rats. Br J Nutr 2008;99:1013–7.
[43] Takemura N, Okubo T, Sonoyama K. *Lactobacillus plantarum* strain No. 14 reduces adipocyte size in mice fed high-fat diet. Exp Biol Med 2010;235:849–56.
[44] Schaafsma G, Meuling WJ, van Dokkum W, Bouley C. Effects of a milk product, fermented by *Lactobacillus acidophilus* and with fructo-oligosaccharides added, on blood lipids in male volunteers. Eur J Clin Nutr 1998;52:436–40.
[45] Ooi LG, Ahmad R, Yuen KH, Liong MT. *Lactobacillus gasseri* [corrected] CHO-220 and inulin reduced plasma total cholesterol and low-density lipoprotein cholesterol via alteration of lipid transporters. J Dairy Sci 2010;93:5048–58.
[46] Greany KA, Bonorden MJ, Hamilton-Reeves JM, McMullen MH, Wangen KE, Phipps WR, et al. Probiotic capsules do not lower plasma lipids in young women and men. Eur J Clin Nutr 2008;62:232–7.
[47] Kiessling G, Schneider J, Jahreis G. Long-term consumption of fermented dairy products over 6 months increases HDL cholesterol. Eur J Clin Nutr 2002;56:843–9.
[48] Gibson GR, McCartney AL. Modification of the gut flora by dietary means. Biochem Soc Trans 1998;26:222–8.
[49] Schneeman BO. Fiber, inulin and oligofructose: similarities and differences. J Nutr 1999;129:1424S–7S.
[50] Simons LA, Gayst S, Balasubramaniam S, Ruys J. Long-term treatment of hypercholesterolaemia with a new palatable formulation of guar gum. Atherosclerosis 1982;45:101–8.
[51] Levrat-Verny MA, Behr S, Mustad V, Remesy C, Demigne C. Low levels of viscous hydrocolloids lower plasma cholesterol in rats primarily by impairing cholesterol absorption. J Nutr 2000;130:243–8.
[52] Andersson H. The ileostomy model for the study of carbohydrate digestion and carbohydrate effects on sterol excretion in man. Eur J Clin Nutr 1992;46(Suppl. 2):S69–76.
[53] Minekus M, Jelier M, Xiao JZ, Kondo S, Iwatsuki K, Kokubo S, et al. Effect of partially hydrolyzed guar gum (PHGG) on the bioaccessibility of fat and cholesterol. Biosci Biotechnol Biochem 2005;69:932–8.
[54] Marlett JA, Hosig KB, Vollendorf NW, Shinnick FL, Haack VS, Story JA. Mechanism of serum cholesterol reduction by oat bran. Hepatology 1994;20:1450–7.
[55] van Munster IP, Tangerman A, Nagengast FM. Effect of resistant starch on colonic fermentation, bile acid metabolism, and mucosal proliferation. Dig Dis Sci 1994;39:834–42.
[56] van Bennekum AM, Nguyen DV, Schulthess G, Hauser H, Phillips MC. Mechanisms of cholesterol-lowering effects of dietary insoluble fibres: relationships with intestinal and hepatic cholesterol parameters. Br J Nutr 2005;94:331–7.
[57] Moundras C, Behr SR, Remesy C, Demigne C. Fecal losses of sterols and bile acids induced by feeding rats guar gum are due to greater pool size and liver bile acid secretion. J Nutr 1997;127:1068–76.

[58] Delzenne NM, Kok N. Effect of non-digestible fermentable carbohydrates on hepatic fatty acid metabolism. Biochem Soc Trans 1998;26:228–30.
[59] Knapper JME, Puddicombe SM, Morgan LM, Fletcher JM. Investigations into the actions of glucose-dependent insulinotropic polypeptide and glucagon-like peptide-1(7-36)amide on lipoprotein-lipase activity in explants of rat adipose-tissue. J Nutr 1995;125:183–8.
[60] Arora T, Loo RL, Anastasovska J, Gibson GR, Tuohy KM, Sharma RK, et al. Differential effects of two fermentable carbohydrates on central appetite regulation and body composition. PLoS One 2012;7:e43263.
[61] Bergman EN. Energy contributions of volatile fatty-acids from the gastrointestinal-tract in various species. Physiol Rev 1990;70:567–90.
[62] Demigne C, Morand C, Levrat MA, Besson C, Moundras C, Remesy C. Effect of propionate on fatty-acid and cholesterol-synthesis and on acetate metabolism in isolated rat hepatocytes. Br J Nutr 1995;74:209–19.
[63] Wolever TMS, Brighenti F, Royall D, Jenkins AL, Jenkins DJA. Effect of rectal infusion of short chain fatty-acids in human-subjects. Am J Gastroenterol 1989;84:1027–33.
[64] Wolever TMS, Spadafora PJ, Cunnane SC, Pencharz PB. Propionate inhibits incorporation of colonic [1,2-C-13]acetate into plasma-lipids in humans. Am J Clin Nutr 1995;61:1241–7.
[65] Singh V, Chassaing B, Zhang L, San Yeoh B, Xiao X, Kumar M, et al. Microbiota-dependent hepatic lipogenesis mediated by stearoyl CoA desaturase 1 (SCD1) promotes metabolic syndrome in TLR5-deficient mice. Cell Metab 2015;22:983–96.
[66] Tolhurst G, Heffron H, Lam YS, Parker HE, Habib AM, Diakogiannaki E, et al. Short-chain fatty acids stimulate glucagon-like peptide-1 secretion via the G-protein-coupled receptor FFAR2. Diabetes 2012;61:364–71.
[67] Samuel BS, Shaito A, Motoike T, Rey FE, Backhed F, Manchester JK, et al. Effects of the gut microbiota on host adiposity are modulated by the short-chain fatty-acid binding G protein-coupled receptor, Gpr41. Proc Natl Acad Sci USA 2008;105:16767–72.
[68] Lin HV, Frassetto A, Kowalik EJ Jr, Nawrocki AR, Lu MM, Kosinski JR, et al. Butyrate and propionate protect against diet-induced obesity and regulate gut hormones via free fatty acid receptor 3-independent mechanisms. PLoS One 2012;7:e35240.
[69] Bjursell M, Admyre T, Goransson M, Marley AE, Smith DM, Oscarsson J, et al. Improved glucose control and reduced body fat mass in free fatty acid receptor 2-deficient mice fed a high-fat diet. Am J Physiol Endocrinol Metab 2011;300:E211–20.
[70] Kimura I, Ozawa K, Inoue D, Imamura T, Kimura K, Maeda T, et al. The gut microbiota suppresses insulin-mediated fat accumulation via the short-chain fatty acid receptor GPR43. Nat Commun 2013;4:1829.
[71] Lin HV, Frassetto A, Kowalik EJ, Nawrocki AR, Lu MFM, Kosinski JR, et al. Butyrate and propionate protect against diet-induced obesity and regulate gut hormones via free fatty acid receptor 3-independent mechanisms. PLoS One 2012;7:7.
[72] Boatman PD, Lauring B, Schrader TO, Kasem M, Johnson BR, Skinner P, et al. (1a*R*,5a*R*)1a,3,5,5a-Tetrahydro-1*H*-2,3-diaza-cyclopropa[*a*]pentalene-4-carboxylic acid (MK-1903): a potent GPR109a agonist that lowers free fatty acids in humans. J Med Chem 2012;55:3644–66.
[73] Tunaru S, Kero J, Schaub A, Wufka C, Blaukat A, Pfeffer K, et al. PUMA-G and HM74 are receptors for nicotinic acid and mediate its anti-lipolytic effect. Nat Med 2003;9:352–5.
[74] Lukasova M, Malaval C, Gille A, Kero J, Offermanns S. Nicotinic acid inhibits progression of atherosclerosis in mice through its receptor GPR109A expressed by immune cells. J Clin Invest 2011;121:1163–73.
[75] Wostmann BS, Wiech NL, Kung E. Catabolism and elimination of cholesterol in germ-free rats. J Lipid Res 1966;7:77–82.

[76] Wostmann BS. Intestinal bile acids and cholesterol absorption in the germfree rat. J Nutr 1973;103:982–90.
[77] Huang Y, Wang JF, Quan GH, Wang XJ, Yang LF, Zhong LL. *Lactobacillus acidophilus* ATCC 4356 prevents atherosclerosis via inhibition of intestinal cholesterol absorption in apolipoprotein E-knockout mice. Appl Environ Microbiol 2014;80:7496–504.
[78] Usman, Hosono A. Bile tolerance, taurocholate deconjugation, and binding of cholesterol by *Lactobacillus gasseri* strains. J Dairy Sci 1999;82:243–8.
[79] Kimoto H, Ohmomo S, Okamoto T. Cholesterol removal from media by lactococci. J Dairy Sci 2002;85:3182–8.
[80] Hamad EM, Sato M, Uzu K, Yoshida T, Higashi S, Kawakami H, et al. Milk fermented by *Lactobacillus gasseri* SBT2055 influences adipocyte size via inhibition of dietary fat absorption in Zucker rats. Br J Nutr 2009;101:716–24.
[81] Lye HS, Rusul G, Liong MT. Removal of cholesterol by lactobacilli via incorporation and conversion to coprostanol. J Dairy Sci 2010;93:1383–92.
[82] Roche HM, Noone E, Sewter C, Mc Bennett S, Savage D, Gibney MJ, et al. Isomer-dependent metabolic effects of conjugated linoleic acid: insights from molecular markers sterol regulatory element-binding protein-1c and LXRalpha. Diabetes 2002;51:2037–44.
[83] Takahashi Y, Kushiro M, Shinohara K, Ide T. Dietary conjugated linoleic acid reduces body fat mass and affects gene expression of proteins regulating energy metabolism in mice. Comp Biochem Physiol B Biochem Mol Biol 2002;133:395–404.
[84] Ho JN, Choi JW, Lim WC, Kim MK, Lee IY, Cho HY. Kefir inhibits 3T3-L1 adipocyte differentiation through down-regulation of adipogenic transcription factor expression. J Sci Food Agric 2013;93:485–90.
[85] Ji YS, Kim HN, Park HJ, Lee JE, Yeo SY, Yang JS, et al. Modulation of the murine microbiome with a concomitant anti-obesity effect by *Lactobacillus rhamnosus* GG and *Lactobacillus sakei* NR28. Benef Microbes 2012;3:13–22.
[86] Aronsson L, Huang Y, Parini P, Korach-Andre M, Hakansson J, Gustafsson JA, et al. Decreased fat storage by *Lactobacillus paracasei* is associated with increased levels of angiopoietin-like 4 protein (ANGPTL4). PLoS One 2010;5:5.
[87] Fak F, Backhed F. *Lactobacillus reuteri* prevents diet-induced obesity, but not atherosclerosis, in a strain dependent fashion in Apoe$^{-/-}$ mice. PLoS One 2012;7:e46837.
[88] Ma X, Hua J, Li Z. Probiotics improve high fat diet-induced hepatic steatosis and insulin resistance by increasing hepatic NKT cells. J Hepatol 2008;49:821–30.
[89] Ley RE, Turnbaugh PJ, Klein S, Gordon JI. Microbial ecology: human gut microbes associated with obesity. Nature 2006;444:1022–3.
[90] Turnbaugh PJ. Microbiology: fat, bile and gut microbes. Nature 2012;487:47–8.
[91] Angelakis E, Bastelica D, Ben Amara A, El Filali A, Dutour A, Mege JL, et al. An evaluation of the effects of *Lactobacillus ingluviei* on body weight, the intestinal microbiome and metabolism in mice. Microb Pathog 2012;52:61–8.
[92] Schwiertz A, Taras D, Schafer K, Beijer S, Bos NA, Donus C, et al. Microbiota and SCFA in lean and overweight healthy subjects. Obesity 2010;18:190–5.
[93] Seo M, Inoue I, Tanaka M, Matsuda N, Nakano T, Awata T, et al. *Clostridium butyricum* MIYAIRI 588 improves high-fat diet-induced non-alcoholic fatty liver disease in rats. Dig Dis Sci 2013;58:3534–44.
[94] Endo H, Niioka M, Kobayashi N, Tanaka M, Watanabe T. Butyrate-producing probiotics reduce nonalcoholic fatty liver disease progression in rats: new insight into the probiotics for the gut-liver axis. PLoS One 2013;8:e63388.
[95] Roberfroid MB. Introducing inulin-type fructans. Br J Nutr 2005;93:S13–25.

CHAPTER 6

Dietary Fiber and Risk of Cardiovascular Diseases

Divya R. Gunashekar, Ram B. Singh, Mohammad A. Niaz, Anand R. Shewale, Toru Takahashi, Anil K. Chauhan, Ravi P. Singh

1 INTRODUCTION

The decline in the morbidity and mortality due to cardiovascular disease (CVD) and coronary artery disease (CAD) in North America and in some European countries has been attributed to changes in diet and lifestyle [1–4]. The increase in morbidity and mortality due to CVDs in developing countries is also due to a rapid increase in the consumption of unhealthy diets and decrease in the functional food consumption. The rates of CVD have long since peaked for many developed countries and mortality from the disease is decreasing. However, the mortality still accounts for almost half (48%) of all deaths in Europe and a third (32.8%) of all deaths in the United States [1–4]. The Global Burden of Disease study revealed that global life expectancy for both sexes increased from 65.3 years in 1990, to 71.5 years in 2013 [1]. However, the number of deaths increased from 47.5 million to 54.9 million during the same period. The absolute and relative differences increased for women aged 25–39 years and older than 75 years and for men aged 20–49 years and 65 years and older [1]. The general pattern of reductions in age-sex specific mortality has been associated with a progressive shift toward a larger share of the remaining deaths caused by noncommunicable disease (NCD) and injuries [1].

The age-standardized death rates for seven substantial causes are increasing, suggesting the potential for reversals in some countries. However, some gaps exist in the empirical data for cause of mortality and morbidity estimates for some countries, because only regional data for India are available for the past decade [5–8]. In an epidemiological study from India, death records of 2222 (1385 men and 837 women) decedents, aged 25–64 years, out of 3034 death records, were randomly selected and studied by verbal autopsy questionnaires [5,7]. Circulatory diseases as the cause of deaths were observed among 29.1% (n = 646) of decedents, including heart attacks

Dietary Fiber for the Prevention of Cardiovascular Disease
http://dx.doi.org/10.1016/B978-0-12-805130-6.00006-9

Both sexes, age-standardized, 2013, deaths per 100,000

	China	India	US	Indonesia	Brazil	Pakistan	Nigeria	Bangladesh	Russia	Japan
High blood pressure	1	1	1	1	1	1	1	1	1	1
High sodium	2	2	6	12	4	3	7	4	6	2
Low fruit	3	4	7	3	5	4	5	2	5	4
Smoking	4	5	9	2	7	6	13	6	4	5
Ambient particulate matter	5	7	16	10	13	9	6	7	14	10
High body-mass index	6	11	2	7	2	2	2	14	2	7
Low whole grains	7	13	11	5	10	15	12	8	7	9
High fasting plasma glucose	8	8	4	11	8	10	11	10	9	11
Household air pollution	9	6		8	20	8	3	3		
Lead	10	18	19	18	19	17	17	13	21	19
Low physical activity	11	9	5	9	6	13	8	9	11	6
High total cholesterol	12	3	3	6	3	5	9	12	3	3
Low omega-3	13	12	12	22	14	14	14	16	17	21
Low fiber	14	15	15	15	16	18	19	18	16	14
Low nuts and seeds	15	16	10	13	12	16	16	17	12	12
Alcohol use	16	20	20	19	21	22	15	21	8	16
Low vegetables	17	10	8	4	9	7	10	5	10	8
Low glomerular filtration	18	14	14	14	11	12	4	11	15	13
Low PUFA	19	17	18	16	18	19	18	19	18	15
Secondhand smoke	20	22	22	17	22	21	22	22	19	18
High trans fat	21	19	17	21	15	11	20	15	20	17
High processed meat	22	21	13	20	17	20	21	20	13	20
High sweetened beverages	23	23	21	23	23	23	23	23	22	22

Figure 6.1 ***Dietary factors and other risk factors in relation to mortality in the global burden of disease study.*** *(Modified from GBD 2013 Mortality and Causes of Death Collaborators. Global, regional, and national age-sex specific all-cause and cause-specific mortality for 240 causes of death, 1990–2013: a systematic analysis for the Global Burden of Disease Study 2013. Lancet 2015; 385: 117–71).*

(10%), strokes (7.8%), valvular heart disease (7.2%, $n = 160$), sudden cardiac death, and inflammatory cardiac diseases (each 2.0%, $n = 44$). Diabetes mellitus as the cause of death was noted among 2.2% ($n = 49$) of decedents, which were mainly due to vascular diseases. Thus CVDs as the cause of deaths were observed among 31.3% of decedents in north Indian urban area [5]. The risk factors for the deaths due to CVDs were unhealthy diet characterized by increased intake of refined foods, such as syrups, biscuits, cakes, bread, refined rice, potato, and so on in conjunction with tobacco, alcoholism, and sedentary behavior [7–11] (Fig. 6.1). Recent guidelines have also emphasized about the adverse effects of refined and sugary foods and beneficial effects of fiber-rich foods [12–14]. There was a marked alteration in behavioral risk factors to biological risk factors in the development of CVDs (Fig. 6.2). The evolutionary diet of *Homo sapiens* descended from

"Heidelberg man" that appeared around 600,000 years ago who lived in the regions of Africa, Europe, and West Asia also consumed foods rich in high fiber, antioxidants, vitamins, minerals, as well as essential and nonessential amino acids (Fig. 6.1). The aim of the present review is to evaluate the recent literature, in populations, concerning dietary fiber intake and risk of CVDs and to update reports published by the International College of Nutrition (Figs. 6.3–6.6).

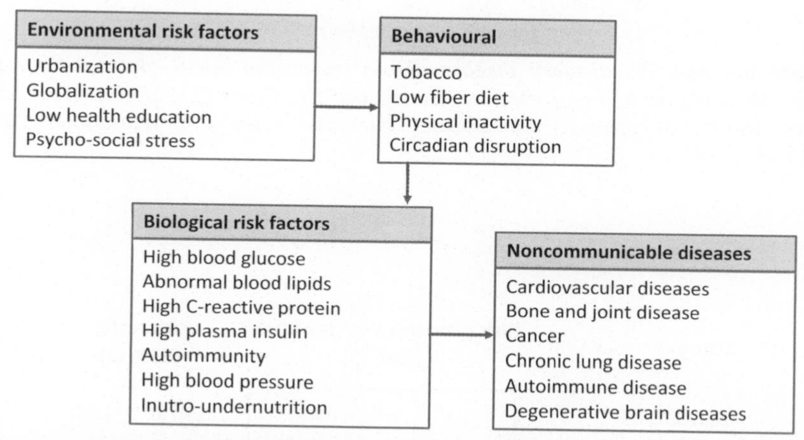

Figure 6.2 ***Pathway for development of cardiovascular disease due to dietary changes.***

	–300 million y	–4 million y	–10 thousand y	1950	2000
Total fat	25%	25%	25%	33%	40%
P:S	1:1	1:1	0.9:1	0.75:1	0.5:1
n–3:*n*–6	1:1	1:1		10:1	20:1
Amino acids	33%	33%	16%–20%	14%	12%
Polyphenols	++++	+++	+++	++	+
Inflammatory index	–/+	–/+		+	+
Chronic diseases	Absent	Absent		Epidemic	
Blood cholesterol	Neutral	Neutral		Atherogenic	
Diet		**Greens + Game**		**Grains + Livestock**	

Figure 6.3 ***Dietary changes during evolution from Heidelberg man to modern man.*** *(From Peressini T., et al. Societal determinants of health and nutrition. Open Access Scientific Reports. 2012;1:513).*

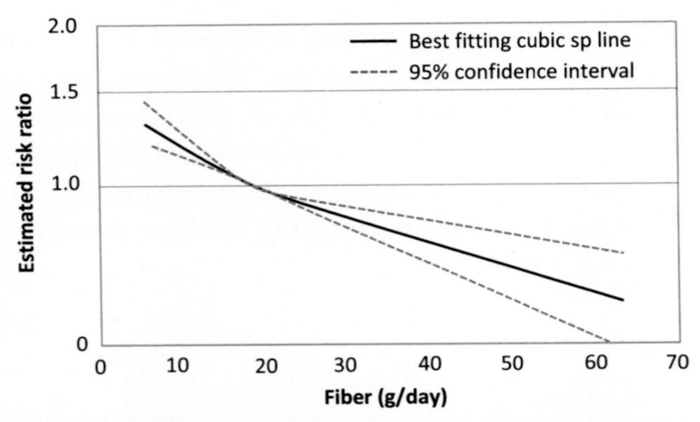

Figure 6.4 ***Risk of coronary disease across increasing levels of total fiber intake.*** *(Modified from Threapleton DE, Greenwood DC, Evans CEL, et al. Dietary fibre intake and risk of cardiovascular disease: systematic review and meta-analysis. BMJ 2013;347:f6879).*

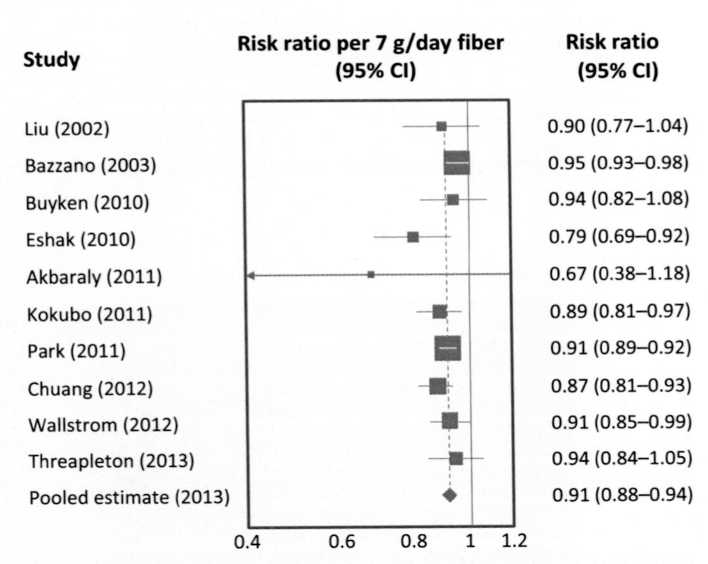

Figure 6.5 ***Risk of cardiovascular diseases associated with each 7 g/day increase in total fiber intake.*** *(Modified from Threapleton DE, Greenwood DC, Evans CEL, et al. Dietary fibre intake and risk of cardiovascular disease: systematic review and meta-analysis. BMJ 2013;347:f6879).*

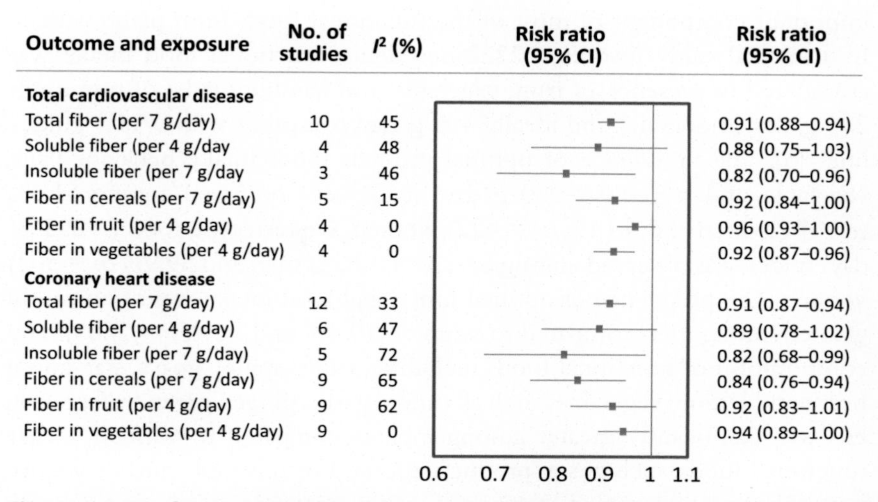

Figure 6.6 ***Combined risk estimates for coronary disease and cardiovascular disease associated with greater intake of total fiber, soluble or insoluble fiber and fiber from different food sources.*** *(Modified from Threapleton DE, Greenwood DC, Evans CEL, et al. Dietary fibre intake and risk of cardiovascular disease: systematic review and meta-analysis. BMJ 2013;347:f6879).*

2 DIETARY FIBER AND RISK OF CARDIOVASCULAR DISEASES

The dietary fiber consumption should be approximately 25 g/day in women and 30–35 g/day among men depending upon body weight but the average person consumes hardly 10–20 g/day depending upon the content of foods. Eating fiber-rich whole foods is the best way to increase fiber intake and the major sources of fiber in the diets are vegetables, fruits, whole grains, nuts, roots, and tubers. Oats and psyllium are rich sources of fiber but randomized controlled clinical trials reported that they have nonsignificant effects on blood pressure and blood lipids.

Evidence from prospective studies is consistent in showing an inverse association between dietary fiber intake and risk of CAD [15,16]. However, it is not clear that dietary fiber from various food sources differs in their effect on CAD risk [15,16]. In the earlier studies, the protective link was proposed between dietary fiber in the form of whole grain foods and CAD. In several cohort studies, the relationship between dietary fiber or fiber-rich foods and total CVDs risk or cardiovascular risk factorsuch as: hypertension, central obesity, diabetes, insulin sensitivity, oxidative stress, inflammation, and elevated plasma cholesterol has been examined. Dietary fiber is an important component of most of the functional foods from plant sources. In a clinical study involving 2222 decedents, functional food intake was considered in presence of fruit, vegetable, and legume intake of minimum 250 g/day (moderate) and ideally 400 g/day consistent with WHO guidelines [5]. The prevalence of optimal prudent foods intake behavior fruit, vegetable, and legume (>250 g/day) intake was observed among 51.4% (n = 712) men and 50.4% (n = 422) women. Western type food (>255 g/day) intake was observed among 63.2% (n = 875) men and 59.9% (n = 502) women. The prevalence of optimal functional food intake was significantly greater among men compared to women (19.4% vs. 14.6%, $P < 0.05$). The consumption of functional foods including, fruits: apples, guava, star-goose berry, and lemon; vegetables such as: onion, garlic, ginger, green leaf, gourds, etc. was significantly greater among men as compared to women. Other functional foods, such as soy products, mustard or olive oil, curd or yogurt, nuts fish, tea cocoa, spices, turmeric, cumin, coriander seeds, and peppers intake showed no significant difference between two sexes [5]. After adjustment of age and body weight, total functional foods intakes and fruit, vegetable, legume, and nuts intake were significantly inversely associated with deaths due to NCDs, whereas Western-type foods (red meat and eggs, refined foods, sugary foods) were positively associated with these causes of deaths in both sexes. Total spices intake, mustard/olive oil intake, and curd or yogurt intake, were inversely but weakly ($P < 0.05$) associated with causes of deaths due to NCDs, among both men and women. It is likely that the consumption of functional foods appears to be lower among victims dying due to CVDs and cancer.

The European Prospective Investigation into Cancer and Nutrition-Heart study, involving 306,331 men and women from 8 European countries, examined dietary fiber intake using center or country-specific diet questionnaires [16]. After a mean follow-up of 11.5 years, there were 2381 deaths due to CAD among subjects without CVDs at baseline. The calibrated intake of dietary fiber was inversely associated with CAD mortality. Each 10 g/day intake of fiber was associated with a 15% lower risk [relative risk (RR) 0.85; 95% confidence interval (CI): 0.73–0.99, P = 0.031]. There was no difference in the associations of the individual food sources of dietary fiber with the risk of CAD mortality but other sources of fiber were less than 5 g/day, which was too low to have therapeutic effects. It is proposed that higher consumption of dietary fiber may cause a lower risk of fatal CAD with no clear difference in the association with CAD for fiber from cereals, fruits or vegetables [16]. A systematic review of 22 cohort studies examined dose-response metaanalysis of cohort studies in relation to CVDs [15]. Total dietary consumption of fiber, the subtypes of fibers, or fiber from food sources in relation to primary events of CVDs including CAD were found out. The metaanalysis revealed that the total dietary fiber intake was inversely associated with risk of CVDs [risk ratio 0.91 per 7 g/day (95% CIs 0.88–0.94)], as well as with CAD [0.91 (0.87–0.94)]. There was also an independent inverse association of intake of insoluble fiber and fiber from cereal and vegetable sources, with the risk of CVDs, as well as with CAD. Only fruit fiber intake was inversely associated with risk of CVDs without any association independently with CAD [15]. It is interesting that every additional 7 g/day of total fiber intake, was associated with a significantly lower risk of 9% for both CVD and CAD [15]. It seems that most European populations were adapted to consume lower dietary fiber; hence even modest increase in dietary fiber may be able to be protective [15–17]. The dietary fiber intake is two-fold greater in Asians compared to Europeans except for Mediterranean countries, which have higher fiber consumption [17–22].

In the United Kingdom Women's Cohort Study a positive association between dietary fiber intake and CVDs was reported. This cohort consisted of 31,036 women, [18]. All subjects were free from history of CVD at baseline with mean age 51.8 years (standard deviation 9.2). After 14.3 years, there were 258 fatal CVD cases (130 stroke, 128 CAD). Total dietary fiber and fiber from different food sources were not associated with fatal CAD, stroke, or CVD risk in the full sample. However, for every 6 g/day increase in NSP, the hazard ratio (HR) was 0.91 (95% CI 0.76–1.08) or for every 11 g/day increase in fiber assessed as AOAC, the HR was 0.92 (95% CI 0.80–1.05). There was also a protective association between cereal sources of fiber on fatal stroke risk in overweight women, HR 0.80 (95% CI 0.65–0.93) $P < 0.01$. It is possible that all studies do not support the inverse association of fiber intake with CVDs [18]. However, a sample of health-conscious women with greater dietary fiber intake may confer no additional cardiovascular benefit, in terms of mortality, but may contribute to lower fatal stroke risk in some subgroups, such as overweight women.

The Swedish population-based Malmö Diet and Cancer cohort examined data from 8,139 male and 12,535 female participants (aged 44–73 years) [19]. The participants were without the history of CVD and diabetes mellitus, and had reported stable dietary habits in the study questionnaire. After a follow-up of 13.5 years, 1089 male and 687 female patients suffered from CVD. High fiber intakes were associated with lower incidence rates of

CVD in women and of ischemic stroke in men. In posthoc analysis, there was a significant interaction between intake of fiber and saturated fat; these interactions also differed between men and women ($P < 0.001$). A high fiber intake was associated with lower risk of CVD, but there were no robust associations between other macronutrients and CVD risk [19]. In a nested case-control study involving 25,639 men and women aged 40–79 years, surveyed from 1993 to 2007, included 2,151 cases with CAD and 5,354 controls for analysis [20]. Among men, age-adjusted CAD risk was inversely associated with 7DD fiber [odds ratio (OR) 0.84, 95% CI 0.79–0.90], but not with FFQ fiber (OR 0.96, 95% CI 0.90–1.12). Among women, age-adjusted CHD risk was inversely associated with 7DD fiber (OR 0.83, 95% CI 0.75–0.93), and had a weaker inverse borderline significant association with FFQ fiber (OR 0.93, 95% CI 0.87–1.01). It is possible that inconsistencies in the diet-CAD relationships in population studies may be associated with the use of different dietary assessment methods [20]. The Nurses' Health Study used competing risks survival analysis to evaluate associations of lifestyle and dietary factors with all-cause and cause-specific mortality among 50,112 participants [21]. There were 4,893 deaths between 1986 and 2004: 1,026 from CVDs, 931 from smoking-related cancers, 1,430 from cancers not related to smoking, and 1,506 from all other causes. Age, body mass index (BMI) at age 18 years, weight change, height, current smoking and pack-years of smoking, glycemic load, cholesterol intake, systolic blood pressure and use of blood pressure medications, diabetes, parental myocardial infarction before age 60 years, and time since menopause were directly related to all-cause mortality, whereas there were inverse associations for physical activity and intakes of nuts, polyunsaturated fat, and cereal fiber. A further meta-analysis, systematically examined investigating whole-grain and fiber intake in relation to risk of type 2 diabetes, CVD, weight gain, and metabolic risk factors [22]. The analysis included 45 prospective cohort studies and 21 randomized-controlled trials between 1966 and February 2012 [22]. The authors reported that compared with never/rare consumers of whole grains, those consuming 48–80 g whole grain per day (3–5 serving/day) had an ~26% lower risk of diabetes [RR = 0.74 (95% CI: 0.69, 0.80)], 21% lower risk of CVD [RR = 0.79 (95% CI: 0.74, 0.85)], and consistently less weight gain during 8–13 years (1.27 vs. 1.64 kg; $P = 0.001$). In the randomized trials, weighted mean differences in postintervention circulating concentrations of fasting glucose and total, and low-density lipoprotein (LDL)–cholesterol comparing whole-grain intervention groups with controls indicated significantly lower concentrations after whole-grain interventions [differences in fasting glucose: −0.93 mmol/L (95% CI: −1.65, −0.21), total cholesterol: −0.83 mmol/L (−1.23, −0.42); and LDL-cholesterol: −0.82 mmol/L (−1.31, −0.33)] (corrected). This metaanalysis provides evidence to support beneficial effects of whole-grain intake on vascular disease prevention [22]. However, potential mechanisms responsible for whole grains' effects on metabolic intermediates require further investigation in large intervention trials. The role of fiber in the etiology of CVDs has been reported in various studies from Asia [23–27].

The Zutphen Study involving a cohort of 1373 men showed that over a follow-up of 40 years, 1130 men died, 348 as a result of CAD [28]. The findings showed that every additional 10 g of recent dietary fiber intake per day reduced CAD mortality by 17% (95% CI: 2%, 30%) and all-cause mortality by 9% (0%, 18%). The strength of the association between long-term dietary fiber intake and all-cause mortality decreased from age 50 years (HR: 0.71; 95% CI: 0.55, 0.93) until age 80 years (0.99; 0.87, 1.12). The strength of the association between dietary fiber and all-cause mortality decreased with increasing age [28]. In another cohort study among 7319 participants (mean age: 49.5 years; range: 39–63 years; 30.3% women) from the Whitehall II Study, alternate healthy eating index was made based on intake of 9 components: vegetables, fruit, nuts, and soy, white or red meat, trans fat, polyunsaturated or saturated fat, fiber, multivitamin use, and alcohol with mortality risk [29]. After a follow-up of 18 years, in the top compared with the bottom third of the healthy eating index score showed 25% lower all-cause mortality (HR: 0.76; 95% CI: 0.61, 0.95) and >40% lower mortality from CVD (HR: 0.58; 95% CI: 0.37, 0.91). Consumption of nuts and soy and moderate alcohol intake appeared to be the most important independent contributors to decreased mortality risk. The encouragement of adherence to the healthy dietary recommendations constitutes a valid and clear public health recommendation that would decrease the risk of premature death from CVD [29].

In a large European prospective study of 452,717 men and women, a total of 23,582 deaths were recorded during a mean follow-up of 12.7 years [30]. Fiber intake was inversely associated with total mortality (HR/per 10 g/day increase): 0.90; 95% CI: 0.88, 0.92); with mortality from circulatory (HR/per 10 g/day increase): 0.90 and 0.88 for men and women, respectively), digestive (HR: 0.61 and 0.64), respiratory (HR: 0.77 and 0.62), and non-CVD noncancer inflammatory (HR: 0.85 and 0.80) diseases; and with smoking-related cancers (HR: 0.86 and 0.89) but not with nonsmoking–related cancers (HR: 1.05 and 0.97) [30]. The associations were more evident for fiber from cereals and vegetables than from fruit. The associations were similar across BMI and physical activity categories but were stronger in smokers and participants who consumed >18 g alcohol/day. This analysis demonstrated that a higher fiber intake is associated with lower mortality, particularly from circulatory, digestive, and non-CVD noncancer inflammatory diseases. In a cohort of 9776 adults participating in the National Health and Nutrition Examination Surveywho were free of CVD at baseline [31]; after a follow-up of 19 years, 1843 incident cases of CAD and 3762 incident cases of CVD were documented. Compared with the lowest quartile of dietary fiber intake (median, 5.9 g/day), participants in the highest quartile (median, 20.7 g/day) had an adjusted RR of 0.88 (95% CI, 0.74–1.04; P =0.05 for trend) for CHD events and of 0.89 (95% CI, 0.80–0.99; P = 0.01 for trend) for CVD events. The RRs for those in the highest (median, 5.9 g/day) compared with those in the lowest (median, 0.9 g/day) quartile of water-soluble dietary fiber intake were 0.85 (95% CI, 0.74–0.98; P =0.004 for trend) for CHD events and 0.90 (95% CI, 0.82–0.99; P =0.01 for trend) for CVD events. Increased intake of dietary fiber, particularly water-soluble fiber, reduces the risk of CAD [28–31].

3 DIETARY FIBER INTAKE IN ASIA AND RISK OF CVDs

In earlier studies from Asia, dietary fiber intake was as great as 25–30 g/day, which may have been responsible for lower risk of CVDs and CAD before 1990 [18–25]. High dietary fiber intake was also found to be protective against hypertension and CVDs in India [5,12–24]. Asian diets are rich in carbohydrates (50%–65% en energy/day) and the consumption of total fat varies between 12% and 25% according to levels of modernization of a particular country in Asia. However, the association between carbohydrate intake and risk of CAD has not been fully explored in Asian populations known to have high-carbohydrate diets [18,23]. In a cohort study from Singapore, the association of consumption of total carbohydrates, different types of carbohydrates, and their food sources were examined with CAD mortality in a Chinese population [25].

This Singapore Chinese Health Study revealed that after a follow-up of 15 years, 1,660 deaths due to CAD occurred, during 804,433 person-years of follow-up [25]. Total carbohydrate intake was not associated with mortality due to CAD risk. However, types of carbohydrates, analyzed individually, showed that starch (rice, potato, noodles, refined bread, etc.) intake was associated with higher risk [men: 1.03 (95% CI: 0.99, 1.08); women: 1.08,

(95% CI: 1.02, 1.14)] and fiber intake with lower risk of CAD mortality [men: 0.94 (95% CI: 0.82, 1.08); women: 0.71 (95% CI: 0.60, 0.84)], with stronger associations in women than men (both *P*-interaction < 0.01). Further analyses revealed, that a replacement of one daily serving of rice with one daily serving of noodles was associated with higher risk (difference in HR: 26.11%; 95% CI: 10.98%, 43.30%). However, a lower risk of CAD death was reported on replacing one daily serving of rice with one of vegetables (−23.81%; 95% CI: −33.12%, −13.20%), fruit (−11.94%; 95% CI: −17.49%, −6.00%), or whole-wheat bread (−19.46%; 95% CI: −34.28%, −1.29%). Thus, the total amount of carbohydrates consumed was not substantially associated with CAD mortality in this Asian population. The shifting of food sources of carbohydrates toward a higher consumption of fruit, vegetables, and whole grains was associated with lower risk of death due to CAD [25]. Since carbohydrate intakes are typically higher in Asians compared to Western populations, it is important to understand their impact on CAD risk which may have considerable implications for developing guidelines for prevention of CVDs [15,23–25]. Current evidence on the impact of total carbohydrate intake on CAD has primarily come from cohort studies conducted in Western populations [15,16].

Coronary risk factors; hypertension, diabetes mellitus, hypercholesterolemia, tobacco consumption, and obesity, as well as CAD are major health problems in high income, as well as in middle income economies [15–20]. Hyperglycemia as a component of metabolic syndrome or diabetes mellitus appears to be the major risk factor for vascular disease in diabetes, which is known to cause oxidative stress and inflammation [18–20]. Despite a moderate increase in fat intake and low rates of obesity, the risk of CAD and diabetes is rapidly increasing in most of the lower middle income countries [21–23]. It is a paradox that in some of these countries the increased risk of people to diabetes and CAD, especially at a younger age, is difficult to explain by conventional risk factors [23,24,27]. It is possible that the presence of new risk factors especially higher lipoprotein(a) (Lpa), hyperhomocysteinemia, insulin resistance, low high-density lipoprotein (HDL)–cholesterol, and poor nutrition during fetal life, infancy, and childhood may explain at least in part, the cause of this paradox [19,20]. The prevalence of obesity, central obesity, smoking, physical inactivity, and stress are rapidly increasing in low and middle income populations, due to economic development [12,27]. In high income populations, there is a decrease in tobacco consumption, increase in physical activity and dietary restrictions, due to learning of the message of prevention, resulting into reduction in coronary and stroke mortality [1–3].

Hypertension, (5%–10%), diabetes (3%–5%), and CAD (3%–4%) are very low in the adult, rural populations of India, China, and in the African sub-continent, which has less economic development. However, in urban and immigrant populations of India and China, to developed countries, the prevalence of hypertension (>140/90, 25%–30%), diabetes (6%–18%), and CAD (7%–14%) are significantly higher, than they are in some of the high income populations [18–27]. Mean serum cholesterol (180–200 mg/dL), obesity (5%–8%) and dietary fat intake (25%–30% en/day) are paradoxically not very high and do not explain the cause of increased susceptibility to CAD and diabetes in some South Asian countries [24–27]. The atherogenic effect of lipid-related risk factors and refined starches and sugar appears to be magnified in some populations due to the presence of the above factors and results into CVD and diabetes at a younger age in these countries [25–27]. The INTERHEART study showed that the existing risk factors among South Asians and other populations can explain all of the increased risks of myocardial infarction among these populations [32]. In a further substudy, the INTERHEART study reported that increased consumption of Western-type of foods was associated with myocardial infarction whereas fruit, vegetable, legumes, fish, and poultry intake were protective factors against heart attack [33]. The oriental type diets containing rice, potato, tofu, noodles, wheat flour South Asian bread had neutral effects on risk of myocardial infarction.

The INTERMAP study revealed that dietary fat content was much lower in the Asian diets compared to diets in UK and USA, but the micronutrient, dietary fiber, and protein contents particularly omega-3 fatty acids were significantly higher in the Japanese diets which may be responsible for lower risk of CVDs in some Asians [34]. The limits for intake of saturated fatty acid and sugar for health promotion have been suggested in a Scientific Statement of the International College of Nutrition, indicating that eating excess of saturated fat (>7%) and sugar (>10%) can have adverse effects on risk of CVDs and diabetes [35]. These findings may require modification of the existing American and European guidelines, proposed for prevention of CAD, in high-income populations [20–22]. Recently, US Department of Health and Human Services and US Department of Agriculture, as well as other agencies have lifted the ban on upper limit of fat intake indicating that it is not the total fat intake but the excess of a type of fat, (e.g., saturated, trans fat, and omega-6 fat) may be responsible for the increased risk of CVD [35–40].

There is a magnification of CVD risk by refined high-carbohydrate foods in Asian populations, who derive a large proportion of their energy from refined grains and may experience exacerbated glycemic responses to such foods compared to persons of European ancestry [41]. In a traditional Asian diet, white rice is a major source of carbohydrates, which may be associated with higher risk of type 2 diabetes [42]. Functional food security or designer foods (400–500 g/day) rich in fiber, substitution for proatherogenic foods; in conjunction with moderate physical activity and cessation of tobacco, may be protective against deaths and disability due to CVD and diabetes in most of these countries [43–46].

In a study, 86,387 Japanese subjects (age 45–65 years, without CVD or cancer in 1995 as Cohort I and in 1998 as Cohort II) were included [47]. After 899,141 person-years of follow-up, the incidence of 2553 strokes and 684 cases of CAD were observed. The HRs (95% CIs) of CVD for the third to fifth quintiles of total fiber was 0.79 (0.63–0.99), 0.70 (0.54–0.89), and 0.65 (0.48–0.87) in women, respectively, compared with the lowest quintile. Total fiber intake was inversely associated with the incidence of stroke, either cerebral infarction or intracerebral hemorrhage in women. The results for soluble fiber were weak but for insoluble fiber in women were similar to those for total fiber. Higher total dietary fiber consumption was associated with reduced risk of CVD in Japanese nonsmokers. The association between dietary fiber intake and mortality from CVD were examined in a Japanese population in a prospective study of 58,730 Japanese men and women aged 40–79 years [48]. After a follow-up of 14-years, a total of 2080 CVD deaths (983 strokes, 422 CAD, and 675 other CVD) were documented. Total, insoluble, and soluble dietary fiber intakes were inversely associated with risk of mortality from CAD and total CVD for both men and women. For men, the multivariable HR (95% CI) for CHD in the highest versus the lowest quintiles were 0.81 [(95% CI, 0.61–1.09); P-trend = 0.02], 0.48 [(95% CI, 0.27–0.84); P-trend < 0.001], and 0.71 [(95% CI, 0.41–0.97); P-trend = 0.04] for total, insoluble, and soluble fiber, respectively. The respective HR (95% CI) for women was 0.80 [(95% CI, 0.57–0.97); P-trend = 0.01], 0.49 [(95% CI, 0.27–0.86); P-trend = 0.004], and 0.72 [(95% CI, 0.34–0.99); P-trend = 0.03], respectively. The fiber sources show that intakes of fruit and cereal fibers but not vegetable fiber were inversely associated with risk of mortality from CAD [48]. In conclusion, dietary intakes of fiber, both insoluble and soluble fibers, and especially fruit and cereal fibers, may reduce risk of mortality from CAD.

4 DIETARY FIBER, GLYCEMIC INDEX OF FOODS AND RISK OF CVDs

It is well known that presence of fiber in the foods decreases the rate of absorption of nutrients from the gut. While sugar, syrups, and other refined foods are rapidly absorbed, increased content of fiber, protein, and fat, delays the absorption of nutrients, particularly of sugar from the intestines, which have a definite effect on risk of many diseases [35]. The Omni Heart trial reported that the isoenergetic replacement of carbohydrates with unsaturated fats or proteins can reduce insulin resistance and blood pressure and improve plasma lipoprotein profiles [43,44].

In cohort studies, the associations between long-term carbohydrate intake on CAD and CVD have not been consistent [45,46]. The type of carbohydrates consumed or food sources of carbohydrates can explain some of these discrepancies in findings with regards to total carbohydrate intake and cardiovascular health. As there is a possibility that carbohydrates from low-glycemic index (GI) food sources (GI $\leq$ 57) was not associated with CAD risk [46]. High glycemic index (GI) foods are known to cause rapid surges in blood glucose, on ingestion and have been associated with higher CVD risk [46,49]. It is both, the amount of carbohydrate, as well as GI of diets, which are responsible for high glycemic load. Refined carbohydrates, such as sugar sweetened beverages and refined grain cereals with high GI have been reported to increase the risk of diabetes and CAD [50,51]. However, the intake of whole-grain foods that are rich sources of phytochemicals; dietary fiber, magnesium, tocotrienols, omega-3 fatty acids, and lignans has been associated with lower risk of CVDs [52]. It is possible that refined carbohydrate-rich foods may increase risk of CVDs and diabetes directly by straining the glucose-homeostatic system or indirectly by displacing the consumption of whole-grain foods that may exert adverse effects [50–52]. The glycemic responses to glucose and rice in people of Chinese and European ethnicity reveal that they are more rapidly absorbed in Asians compared to Europeans, possibly due to high carbohydrates diets [53]. In a northern Swedish population-based cohort, low-carbohydrate, high-protein diets was associated with beneficial effects on mortality, which suggested that diets rich in vegetables, fruits, whole grain, nuts, and poultry and fish, such as Mediterranean style diets may be protective [54–56].

5 DIETARY FIBER FROM WHOLE GRAIN AND RISK OF CVDs

Increased intake of whole grains has been associated with lower risk of CAD; however, the effects of various types of whole grains, such as rye, oats, wheat, gram, and so on on CAD risk is not clear [57]. In a prospective study including 54,871 Danish adults aged 50–64 years, of whom 2,329 individuals developed myocardial infarction (13.6 years of follow-up). In the highest quartile of total whole grain intake, lower risks of myocardial infarction were observed between both sexes, than for individuals with intake in the lowest quartile. Rye and oats, but not wheat, were associated with lower myocardial infarction risk in men but significant associations were not observed in women. However, in both sexes, total whole-grain products consumption was associated with significantly lower myocardial infarction risks. Rye bread (in men and women) and oatmeal (in men) were associated with significantly lower risk of myocardial infarction, whereas no significant association was shown for whole-grain bread, crispbread, and wheat. It is clear that whole-grain intake is related to lower risk of myocardial infarction and that the cereals rye and oats might especially hold a protective effect. As mentioned earlier in a metaanalysis, compared with never/rare consumers of whole grains, those consuming 48–80 g whole grain per day (3–5 serving per day) had a 26% lower risk of type 2 diabetes [RR = 0.74 (95% CI: 0.69, 0.80)], 21% lower risk of CVD [RR = 0.79 (95% CI: 0.74, 0.85)], and lower weight gain during 8–13 years (1.27 vs. 1.64 kg; $P = 0.001$) [22].

There is scarcity of evidence on the intake of different types of whole grain in relation to all-cause and cause-specific mortality in a healthy population [58,59]. The National Institutes of Health (NIH)-AARP Diet and Health Study, during an average of 9 years of follow-up, identified 20,126 deaths in men and 11,330 deaths in women [58]. Dietary fiber intake was associated with a significantly lowered risk of total death in both men and women (multivariate RR comparing the highest with the lowest quintile, 0.78 (95% CI, 0.73–0.82; P for trend, <0.001) in men and 0.78 (95% CI, 0.73–0.85; P for trend, <0.001) in women. Dietary fiber intake also lowered the risk of death from CVDs, infectious, and respiratory diseases by 24%–56% in men and by 34%–59% in women [58]. Inverse association between dietary fiber intake and cancer death was observed in men but not in women. Dietary fiber from grains, but not from other sources, was significantly inversely related to total and cause-specific death in both men and women. It seems that dietary fiber may reduce the risk of death from CVDs, infectious, and respiratory diseases.

A large Scandinavian HELGA cohort, including 120,010 cohort members aged 30–64 years from the Norwegian women and Cancer Study, the Northern Sweden Health and Disease Study, and the Danish Diet Cancer and Health Study were examined for whole grain intake in relation to all-cause and cause-specific mortality in a healthy population [59]. A total of 3658 women and 4181 men died during the follow-up, indicating lower all-cause mortality with higher intake of total whole grain products [women: MRR 0·89 (95% CI 0.86, 0.91); men: MRR 0.89 (95% CI 0.86, 0.91) for a doubling of intake]. The intake of breakfast cereals and nonwhite bread was associated with lower mortality. The authors also reported lower all-cause mortality with a total intake of different whole grain types [women: MRR 0.88 (95% CI 0.86, 0.92); men: MRR 0.88 (95% CI 0.86, 0.91) for a doubling of intake], such as oat, rye, and wheat [59]. The associations were found in both women and men and for different causes of deaths. It is possible that higher intake of whole grain products and grain types was associated with lower mortality among participants.

6 MEDITERRANEAN STYLE DIETS, DIETARY FIBER AND RISK OF CVDs

Mediterranean-style diets are characterized by vegetables, whole grains, nuts, fish, poultry, virgin olive oil that are rich sources of dietary fiber, micronutrients, antioxidants and monounsaturated fat, and amino acids in the diets. These diets have a low-GI and have been proven to decrease the risk of CVDs and diabetes in cohort studies, as well as in randomized trials [54–56]. The dietary fiber content can be estimated to vary from 25 g to 35 g/day in Mediterranean-style diets. In a typical Mediterranean diet eating population, the median (percentile 25–75) energy-adjusted intake of fiber was 29.4 g/day (23.9–36.4 g/day) [60]. Major sources of fiber intake were fiber from vegetables (median, 10.1; percentile 25–75, 7.3–13.6 g/day), fiber from fruit (5.0, 3.1–8.5 g/day), fiber from cereals (4.8, 3.6–7.7 g/day), and fiber from legumes (3.7, 2.7–4.7 g/day). The study included 171 cases, with myocardial infarction, whom were less than 80 years of age and sex-matched control subjects showing that three upper quintiles of fiber intake were inversely associated with myocardial infarction [60]. After adjustment for nondietary and dietary confounders, an inverse linear trend was clearly significant, indicating the highest relative reduction of risk (86%) for the fifth quintile (OR = 0.14, 95% CI: 0.03–0.67). An inverse association was also apparent for fruit intake, but not for vegetables or legumes. It seems that a substantial part of the protective effect of the Mediterranean diet on coronary risk might be attributed to a high intake of fiber and fruit. The adverse effects of Western diets are mainly due to refining and high content of sugar, glucose, and fructose [61,62].

Since high-GI of foods have been associated with increased risk of CAD, there is a need to find out if low-GI foods can cause incremental benefits with respect to CAD, among people adhering to the traditional Mediterranean diet [63]. The Greek European Prospective Investigation into Cancer and Nutrition included 20,275 participants free of CVDs, cancer, or diabetes at baseline and without incident diabetes. After a median follow-up of 10.4 years, 417 participants developed CAD, including 162 deaths from the disease. A significant positive association of GI with CAD incidence emerged (HR for the highest vs. the lowest tertile = 1.41, 95% CI: 1.05–1.90). The association with glycemic load was more significant among subjects with higher BMI. A greater adherence to Mediterranean diet with low/moderate glycemic load was associated with lower risk of CAD incidence (HR = 0.61, CI: 0.39–0.95) and mortality (HR = 0.47, 95% CI: 0.23–96). It is possible that a high dietary glycemic load increases the risk of CAD and a high glycemic load diet with suboptimal adherence to the traditional Mediterranean pattern. A low/moderate glycemic load diet could lead to a 40% reduced risk for CAD, and over 50% reduced risk for death from CAD [63]. In a further study involving 1658 individuals, the baseline CV risk was estimated with the Framingham risk score [64]. Participants were divided into two groups: individuals at low risk (CV < 10) and individuals with CV risk ≥10. After a 12-year mean follow-up, 220 deaths, 84 due to CV diseases, and 125 incident CV events occurred [64]. The adherence to the Mediterranean diet was low in 768 (score 0–2), medium in 685 (score 4–5), and high in 205 (score >6) individuals. Interestingly the values of BMI, waist circumference, fasting glucose, and insulin significantly decreased from low to high diet adherence only in participants with CV risk ≥10. Greater adherence to the Mediterranean diet was associated with reduced fatal and nonfatal CV events, especially in individuals at low CV risk, thus suggesting the usefulness of promoting this nutritional pattern in particular in healthier individuals [64].

7 RANDOMIZED, CONTROLLED TRIALS WITH MEDITERRANEAN FOODS

Evidence regarding the effect of fruit and vegetable consumption on metabolic syndrome remains inconclusive. In a metaanalysis, 383 articles were identified; including eight randomized controlled trials with 396 participants (205 in intervention groups and 191 in control groups)[65]. Fruit and vegetable intake was associated with a reduction in diastolic blood pressure (standardized mean difference: −0.29; 95% CI: −0.57 to −0.02; $P = 0.04$); however, such intake did not affect waist circumference, systolic blood pressure, fasting glucose, HDL–cholesterol, and triglyceride levels in metabolic syndrome patients. The findings indicate an inverse association between fruit and vegetable consumption and diastolic blood pressure in metabolic syndrome patients [65].

There is evidence that inclusion of high fiber foods, such as oats, fruits, and vegetables in the diet can decrease fat intake and modulate blood lipids and blood pressures. In a clinical trial among 61 group A and 59 group B patients with essential hypertension, guava fruit preferably before meals was administered in a randomized and single-blind fashion for 12 weeks [66]. After 12 weeks, a significant net decrease in serum total cholesterol (9.9%), triglycerides (7.7%), and blood pressures (9.0/8.0 mm Hg) with a significant net increase in HDL–cholesterol (8.0%) after 12 weeks of guava fruit substitution in group A than in group B were recorded. By adding moderate amounts of guava fruit in the usual diet, changes in dietary fatty acids and carbohydrates may occur, providing significant amounts of soluble dietary fiber and antioxidant vitamins and minerals without any adverse effects. There is a greater decrease in lipoprotein levels and blood pressures in the intervention group compared to control group [66]. A randomized, single-blind, controlled trial involving 145 hypertensives that entered the trial, 72 patients were assigned to take 0.5–1.0 kg of guava daily (group A) and 73 patients to low fat NCEP diet (group B) [67], while salt, fat, cholesterol, caffeine, and alcohol intake were similar in both groups. After 4 weeks of follow-up on an increased consumption of dietary potassium and low sodium/potassium ratio, group A patients were associated with 7.5/8.5 mmHg net decrease in mean systolic and diastolic pressures compared with group B. Increased intake of soluble dietary fiber (47.8 ± 11.5 vs. 9.5 ± 0.85 g/day) was associated with a significant decrease in serum total cholesterol (7.9%), triglycerides (7.0%) and an insignificant increase in HDL–cholesterol (4.6%) with a mild increase in the ratio of total cholesterol/HDL–cholesterol in group A patients compared with group B.

In a large controlled trial, fruits, vegetables, and legumes were administered for 12 weeks as an adjunct to a prudent diet in decreasing blood lipids in 310 (intervention; group A) and 311 (control; group B) patients with risk factors for CAD in a parallel, single-blind fashion [68]. Fruits and vegetables decreased total cholesterol level by 6.5% and LDL–cholesterol level by 7.3% in group A, whereas the levels were unchanged in group B. The HDL–cholesterol levels that decreased during the diet stabilization period in both groups, increased by 5.6% in group A after 12 weeks. Serum triglycerides also decreased (7%) more in group A than B. Fasting blood glucose decreased by 6.9% in group A and by 2.6% in group B [68]. Because tasty fruits were taken by the patients before meals (when they are hungry) and are easily available at reasonable cost in our marketing and buying capacity, the compliance was excellent.

Recent results from the PREDIMED studies indicate that Mediterranean-style diets can cause significant decline in CVDs and cancer [69–71]. The beneficial effects of Mediterranean-style diets may be because of increased intake of fruits, vegetables, nuts, fish, poultry, and olive oil with very little red meat[69–71]. Majority of these foods possess increased content of polyphenolic flavonoids, carotenoids, omega-3 fatty acids, antioxidants, vitamins, and minerals, as well as essential and nonessential amino acids [72,73]. Increased consumption of nuts has been demonstrated to cause significant decline in all-cause mortality [6,74], whereas a greater intake of fresh fruits was associated with a significant decline in CVDs in China [75]. Mediterranean-style diets have also been demonstrated to cause significant decline in CVDs in earlier landmark studies [76–79]. Effect of a traditional Mediterranean diet on lipoprotein oxidation showed that there was a nonsignificant reduction in oxidative stress in the intervention group compared to low fat diet group [12,80]. The increase in oxidative stress results in inflammation, which is a unifying hypothesis for predisposing atherosclerosis, carcinogenesis, and osteoporosis, as well as type 2 diabetes [81,82]. In a landmark study among 7447 subjects at high cardiovascular risk, who were randomized to a Mediterranean diet with added olive oil, a Mediterranean diet with added nuts, or a low-fat control group ; after 4.8 years, the risk of combined heart attack, stroke, and death from CVDs was reduced by 30% in the Med + Olive Oil group, and 28% in the Med + Nuts group. The results were only significant in men, not in women and the risk of stroke went down by 39% in the Mediterranean diet groups. There was no statistically significant difference in heart attacks and the dropout rates were twice as high in the control group (11.3%), compared to the Mediterranean diet groups (4.9%). When looking at subgroups, people with high blood pressure, lipid problems, or obesity responded best to the Mediterranean diet. Despite this study being hailed as a success story, there was no statistically significant difference in total mortality (risk of death). A Mediterranean diet with either olive oil or nuts may reduce the combined risk of stroke, heart attack, and death from CVD. There was no statistically significant effect in women and no reduction in mortality.

8 MECHANISMS

Fiber-rich foods, in addition to fiber contain many other potentially beneficial compounds within them, which could have protective effects [77–79]. Protective compounds in grains, such as antioxidants, hormonally active lignans, phytosterols, amylase inhibitors, and saponins have all been shown to influence risk factors for CAD, and the combination of compounds within grains could be responsible for their protective effect. The protective effects of dietary fiber on risk of CVD and CAD are biologically plausible, and there are several potential mechanisms through which fiber can decrease risk factors [65–67]. Soluble, viscous fiber types can affect absorption from the small intestine because of the formation of gels that attenuate postprandial blood glucose and increase in blood lipid rises [79]. The formation of gels also slows gastric emptying, maintaining levels of satiety, and contributing toward less weight gain. Soluble fiber and resistant starch molecules are additionally fermented by bacteria in the large intestine, which produce short-chain fatty acids, that help to reduce circulating cholesterol levels. Dietary fiber and fiber-rich foods can also decrease oxidative stress and inflammation [79].

9 DIETARY FIBER, HIGH GLYCEMIC INDEX FOODS AND GLYCEMIC CONTROL

In many studies, dietary fiber intake was not assessed but the subjects included were receiving increased amount of fiber-rich foods, such as Mediterranean style diets. In a metaanalysis comprising of 3102 studies, the current compilation of human studies was reviewed to determine the association of various doses and durations of fructose consumption on metabolic syndrome [61]. Studies on natural fructose content of foods, nonclinical trials, and trials in which fructose was recommended exclusively as sucrose or high-fructose corn syrup were excluded, and only 15 remaining studies were included in this analysis. Fructose intake was positively associated with increased fasting blood sugar, elevated triglycerides, and elevated systolic blood pressure, which were significant ($P = 0.002$). The corresponding figure was inverse for HDL–cholesterol ($P = 0.001$). It is possible that fructose consumption from industrialized foods has significant effects on most components of metabolic syndrome [61]. Another metaanalysis was conducted to update the evidence on the effect of fructose on CAD lipid targets for CVDs (LDL–cholesterol, apolipoprotein B, non-HDL–cholesterol), and metabolic syndrome (triglycerides and HDL–cholesterol) [62]. Eligibility criteria were met by 51 isocaloric trials ($n = 943$), in which fructose was provided in isocaloric exchange for other carbohydrates, and 8 hypercaloric trials ($n = 125$), in which fructose-supplemented control diets with excess calories compared to the control diets alone without the excess calories. The findings revealed that fructose had no effect on LDL–cholesterol, non-HDL–cholesterol, apolipoprotein B, triglycerides, or HDL–cholesterol in isocaloric trials. Fructose intake increased apolipoprotein B ($n = 2$ trials; mean difference = 0.18 mmol/L; 95% CI: 0.05, 0.30; $P = 0.005$) and triglycerides ($n = 8$ trials; mean difference = 0.26 mmol/L; 95% CI: 0.11, 0.41; $P < 0.001$) in hypercaloric trials [62]. The study is limited by small sample sizes, limited follow-up, and low quality scores of the included trials. The authors concluded that fructose only had an adverse effect on established lipid targets when added to existing diets so as to provide excess calories (+21% to 35% energy). The controversy suggested that further trials that are larger, longer, and of higher quality are required.

The benefits of the Mediterranean diet are known, but its effect on glycemic control has not been totally elucidated. A multicentric parallel trial including 191 participants (77 men and 114 women) of the PREDIMED study compared three dietary interventions: two Mediterranean diets supplemented with virgin olive oil ($n = 67$; BMI = 29.4 ± 2.9) or mixed nuts ($n = 74$; BMI = 30.1 ± 3.1), and a low-fat diet ($n = 50$; BMI = 29.8 ± 2.8) [83]. After follow up of 1 year, there was a rise in adiponectin/leptin ratio ($P = 0.043$, $P = 0.001$ and $P < 0.001$ for low-fat, olive oil, and nut diets, respectively) and adiponectin/HOMA-IR ratio ($P = 0.061$, $P = 0.027$, and $P = 0.069$ for low-fat, olive oil, and nut diets, respectively) with a reduction in waist circumference ($P = 0.003$, $P = 0.001$, and $P = 0.001$ for low-fat, olive oil, and nut diets, respectively) were reported in the three groups. In both Mediterranean diet groups, but not in the low-fat diet group, this was associated with a significant reduction in body weight ($P = 0.347$, $P = 0.003$, and $P = 0.021$ for low-fat, olive oil, and nut diets, respectively). It is possible that Mediterranean diets supplemented with virgin olive oil or nuts reduced total body weight and improved glucose metabolism to the same extent as the usually recommended low-fat diet [83]. A recent metaanalysis, included 20 randomized-controlled trials, which involved ($n = 3073$ included in final analyses across 3460 randomly assigned individuals) [84] the low-carbohydrate, low-GI, Mediterranean, and high-protein diets, all led to a greater improvement in glycemic control [glycated hemoglobin reductions of −0.12% ($P = 0.04$), −0.14% ($P = 0.008$), −0.47% ($P < 0.00001$), and −0.28% ($P < 0.00001$), respectively] were compared with their respective control diets, with the largest effect size seen in the Mediterranean diet. Interestingly, low-carbohydrate and Mediterranean diets led to greater weight loss [−0.69 kg ($P = 0.21$) and −1.84 kg ($P < 0.00001$), respectively], with an increase in HDL seen in all diets except the high-protein diet [84]. It seems that a low-carbohydrate, low-GI, Mediterranean, and high-protein diets are effective in improving various markers of cardiovascular risk in people with diabetes and should be considered in the overall strategy of diabetes management.

In a further metaanalysis, the effects of a Mediterranean diet compared to other dietary interventions on glycemic control irrespective of weight loss were examined [85]. Of 8 studies, which met the inclusion criteria, 7 examined fasting blood glucose ($n = 972$), 6 examined fasting insulin ($n = 1330$), and 3 studies examined HbA1c ($n = 487$). None of the interventions were significantly better than the others in lowering glucose parameters. The Mediterranean diet reduced HbA1c significantly compared to usual care but not compared to the Palaeolithic diet. The findings indicate the need for further research in this area because no firm conclusions about relative effectiveness of interventions could be drawn as a result of the paucity of the evidence. In another metaanalysis, 15 studies met inclusion and exclusion criteria and the overall mean difference of fiber versus placebo was a reduction of fasting blood glucose of 0.85 mmol/L (95% CI, 0.46–1.25) [86]. Dietary fiber as an intervention also had an effect on HbA1c over placebo, with an overall mean difference of a decrease in HbA1c of 0.26% (95% CI, 0.02–0.51). It seems that an intervention involving fiber supplementation for type 2 diabetes mellitus can reduce fasting blood glucose and HbA1c. This suggests that increasing dietary fiber in the diet of patients with type 2 diabetes is beneficial and should be encouraged as a disease management strategy. The risk of CVDs varies according to social classes in Asian countries, which needs to be studied in cohort studies [87–90]. The International College of Nutrition has been active in the recommendations for prevention of CVDs, metabolic syndrome and diabetes by increased intake of dietary fiber [91–95].

10 ECONOMIC BURDEN OF CVDs

The economic burden of CVDs to the society is very high. The United States alone spends more than a billion dollars a day in medical costs and lost productivity due to CVDs and United Kingdom spends about €192 billion a year on treating CVDs [96,97]. The burden is also very high in developing countries. South Africa, for example, spends about one fourth of its total health care expenditure on the direct treatment of CVDs and Brazil spends about one tenth of its total health care expenditure on severe CVDs alone [98,99].

Diet monitoring can help reduce this burden significantly. A study assessing the cost effectiveness of Mediterranean diet showed that even with the additional cost of seeing a dietician and switching to high fiber rich Mediterranean diet as compared to western diet resulted in a cost effectiveness of AU$1013 (US$703, €579) per quality-adjusted year of life (QALY) gained [100]. Compared to the cost effectiveness of a relatively cheap pharmacotherapeutic agent such as aspirin, which has a cost effectiveness of $11,000 per QALY gained, diet monitoring is still the cheaper option [101]. Another study conducted in Canada, projected health care savings ranging from 64.8 million to 1.3 billion Canadian dollars in cardiovascular costs by increasing the uptake of dietary fiber to recommended levels [102]. These studies suggest that diet monitoring is a more cost effective option and results in more cost savings compared to pharmacotherapy in preventing CVDs. It also offers the additional benefit of no side effects as experienced with pharmacotherapy.

One important mechanism of action is that dietary resistant starch stimulates glucagon like peptide-1 and peptides YY with the help of gut microbioda by fermentation of the fiber [103,104]. Fermentation and the liberation of SCFAs in the lower gut are associated with increased proglucagon and PYY gene expression [103,104] (Fig. 6.7). Fermentation of RS is most likely the primary mechanism for increased endogenous secretions of total GLP-1 and PYY in rodents. Thus any factor that affects fermentation should be considered when dietary fermentable fiber is used to stimulate GLP-1 and PYY secretion.

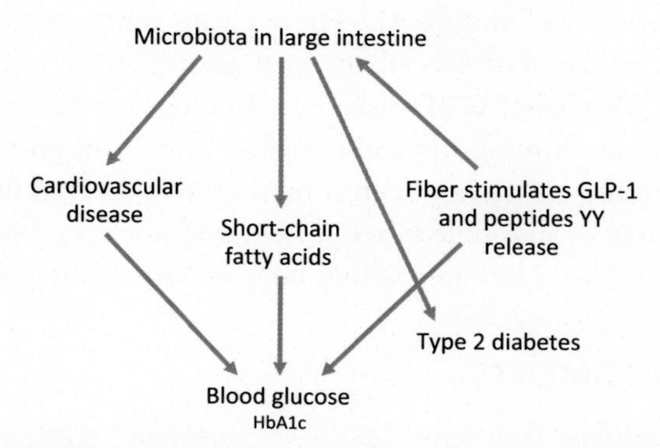

Figure 6.7 ***Mechanism on the effects of fibers on cardiovascular disease and diabetes.***

In conclusion, findings from various studies were aligned with current recommendations to increase fiber intake showing a large risk reduction with an achievable increase in daily fiber intake. Since CVDs are one of the major causes of death in the world and prevalence rates among the population are estimated to be around 13%–16%, a modest decrease in risk could affect many thousands of subjects. It is possible to provide an additional dietary fiber of 7 g via eating only one portion of whole grain plus a portion of beans or lentils, or through two to four servings of fruit and vegetables [12]. Lower risk of CVD and CAD was also observed with greater intakes of insoluble, cereal, fruit, and vegetable fiber.

The findings relate mainly to fiber from food intake rather than from fiber isolates or extracts, and any public health messages musttherefore reflect on total dietary pattern rather than fiber packets. There is a need to find out whether fiber consumedas an extract from certain foods is beneficial needs to be proven by randomized, controlled trials with clinical endpoints or controlled feeding studies with intermediate outcomes. Majority of the studies did not quantify the dose-response association but reported risks for highest compared with lowest consumers. However, recent systematic reviews revealed a lower CVD risk of around 20% for high fiber consumers and separately for high whole grain consumers, compared with the lowest consumers of each. The potential beneficial compounds within fiber-rich foods, which confers the protective associations noted with greater fiber intake, needs identification.

11 SUMMARY

High fiber dietary patterns specifically based on whole grains, vegetable, or fruit sources and rich in insoluble type fiber are significantly associated with lower risk of CVD including CAD. These results reflect recommendations to increase the intake of dietary fiber. Since greater intake of fruit fiber was also associated with lower CVD risk, these findings provide evidence relating to whole food consumption and therefore do not support consumption of foods specifically enriched in cereal or vegetable derived fiber. Evidence relating to soluble or insoluble types of fiber was not very extensive; hence, further investigation in this area would be necessary to establish this fact.

ACKNOWLEDGMENTS

Conflict of interest has not been declared by the authors. Logistic support has been provided by International College of Nutrition and International College of Cardiology.

REFERENCES

[1] GBD 2013 Mortality and Causes of Death Collaborators. Global, regional, and national age-sex specific all-cause and cause-specific mortality for 240 causes of death, 1990–2013: a systematic analysis for the Global Burden of Disease Study 2013. The Lancet 2015;385:117–71.
[2] WHO. Mortality and burden of disease estimates for WHO Member States in 2008. Geneva: World Health Organization; 2010.
[3] Mozaffarian D, Benjamin EJ, Go AS, Arnett DK, Blaha MJ, Cushman M, on behalf of the American Heart Association Statistics Committee and Stroke Statistics Subcommittee. et al. Heart disease and stroke statistics—2015 update: a report from the American Heart Association. Circulation 2016;131:e38–360.
[4] NCD Risk Factor Collaboration. Trends in adult body-mass index in 200 countries from 1975 to 2014: a pooled analysis of 1698 population-based measurement studies with 19·2 million participants. Lancet 2016;387:1377–96.
[5] Singh RB, Visen P, Sharma D, Sharma S, Mondol R, Sharma JP, et al. Study of functional foods consumption patterns among decedents dying due to various causes of death. Open Nutraceut J 2015;8:16–28.
[6] Shehab A, Elkilany G, Singh RB, Hristova K, Chaves H, Cornelissen G, et al. Coronary risk factors in South West Asia. World Heart J 2015;7:21–9.
[7] Singh RB, Fedacko J, Vargova V, Kumar A, Mohan A, et al. Singh's verbal autopsy questionnaire for assessment of causes of death, social autopsy, tobacco autopsy, and dietary autopsy based on medical records and interview. Acta Cardiol 2011;66:471–81.
[8] Pednekar MS, Gupta R, Gupta PC. Illiteracy, low educational status and cardiovascular mortality in India. BMC Public Health 2011;11:567.
[9] Singh RB, Beegom R, Mehta AS, Niaz MA, De AK, et al. Social class, coronary risk factors and undernutrition, a double burden of diseases, in women during transition, in five Indian cities. Int J Cardiol 1999;69:139–47.
[10] Moodie R, Stuckler D, Monteiro C, Sheron N, Thamarangsi, et al. Profits and pandemics: prevention of harmful effects of tobacco, alcohol, and ultra-processed food and drink industries. Lancet 2013;381:670–9.
[11] Singh RB, Takahashi T, Nakaoka T, Otsuka K, Toda E, et al. Nutrition in transition from *Homo sapiens* to Homo economicus. Open Nutra J 2013;6:6–17.
[12] Hristova K, Pella D, Singh RB, Dimitrov BD, Chaves H, Juneja L, et al. Sofia Declaration for Prevention of Cardiovascular Diseases and Type 2 Diabetes Mellitus: a scientific statement of the International College of Cardiology and International College of Nutrition. World Heart J 2014;6:89–106.
[13] Fedacko J, Vargova V, Pella D, Singh RB, Gupta AK, Juneja LR, et al. Sugar and the heart. World Heart J 2014;6:215–8.
[14] MacGregor GA, Hashem KM. Action on sugar—lessons from UK salt reduction programme. Lancet 2014;383:929–31.
[15] Threapleton DE, Greenwood DC, Evans CEL, et al. Dietary fibre intake and risk of cardiovascular disease: systematic review and meta-analysis. BMJ 2013;347:f6879.
[16] Crowe FL, Key TJ, Appleby PN, Overvad K, Schmidt EB, Egeberg R, et al. Dietary fibre intake and ischaemic heart disease mortality: the European Prospective Investigation into Cancer and Nutrition-Heart study. Eur J Clin Nutr 2012;66(8):950–6.
[17] Smolina K, Wright FL, Rayner M, Goldacre MJ. Determinants of the decline in mortality from acute myocardial infarction in England between 2002 and 2010: linked national database study. BMJ 2012;344:d8059.
[18] Threapleton DE, Greenwood DC, Burley VJ, Aldwairji M, Cade JE. Dietary fibre and cardiovascular disease mortality in the UK Women's Cohort Study. Eur J Epidemiol 2013;28:335–46.

[19] Wallstrom P, Sonestedt E, Hlebowicz J, Ericson U, Drake I, Persson M, et al. Dietary fiber and saturated fat intake associations with cardiovascular disease differ by sex in the Malmo diet and cancer cohort: a prospective study. PLoS ONE 2012;7:e31637.

[20] Ward HA, Keogh R, Lentjes M, Luben RN, Wareham NJ, Khaw KT. Fibre intake in relation to serum total cholesterol levels and CHD risk: a comparison of dietary assessment methods. Eur J Clin Nutr 2012;66:296–304.

[21] Baer HJ, Glynn RJ, Hu FB, Hankinson SE, Willett WC, Colditz GA, et al. Risk factors for mortality in the nurses' health study: a competing risks analysis. Am J Epidemiol 2011;173:319–29.

[22] Ye EQ, Chacko SA, Chou EL, Kugizaki M, Liu S. Greater whole-grain intake is associated with lower risk of type 2 diabetes, cardiovascular disease, and weight gain. J Nutr 2012;142:1304–13.

[23] Singh RB, Sircar AR, Rastogi SS, Singh R. Dietary modulators of blood pressure in hypertension. Eur J Clin Nutr 1990;44:319–27.

[24] Singh RB, Beegom R, Verma SP, et al. Association of dietary factors and other coronary risk factors with social class in women in five Indian cities. Asia Pac J Clin Nutr 2000;9:298–302.

[25] Rebello SA, Koh H, Chen C, Naidoo N, Odegaard AO, Koh WP, et al. Amount, type, and sources of carbohydrates in relation to ischemic heart disease mortality in a Chinese population: a prospective cohort study. Am J Clin Nutr 2014;100(1):53–64.

[26] Singh RB, Rastogi V, Niaz MA, Ghosh S, Sy RG, Janus ED. Serum cholesterol and coronary artery disease in populations with low cholesterol levels: the Indian paradox. Int J Cardiol 1998;65:81–90.

[27] Singh RB, Pella D, Kartikey K, DeMeester F. the Five City Study Group. Prevalence of obesity, physical inactivity and undernutrition, a triple burden of diseases, during transition in a middle income country. Acta Cardiol 1997;62:119–27. 28–31.

[28] Streppel MT, Ocke MC, Boshuizen HC, Kok FJ, Kromhout D. Dietary fiber intake in relation to coronary heart disease and all-cause mortality over 40 y: the Zutphen study. Am J Clin Nutr 2008;88:1119–25.

[29] Akbaraly TN, Ferrie JE, Berr C, Brunner EJ, Head J, Marmot MG, et al. Alternative healthy eating index and mortality over 18 y of follow-up: results from the Whitehall II cohort. Am J Clin Nutr 2011;94:247–53.

[30] Chuang S-C, Norat T, Murphy N, Olsen A, Tjønneland A, Overvad K, et al. Fiber intake and total and cause-specific mortality in the European Prospective Investigation into Cancer and Nutrition cohort. Am J Clin Nutr 2012;96:164–74.

[31] Bazzano LA, He J, Ogden LG, Loria CM, Whelton PK. National Health Nutrition Examination Survey. Dietary fiber intake and reduced risk of coronary heart disease in US men and women: the National Health and Nutrition Examination Survey I Epidemiologic Follow-up Study. Arch Int Med 2003;163:1897–904.

[32] Yusuf S, Hawken S, Ounpuu S, Dans T, Avezum A, Lanas F, on behalf of the INTERHEART Study Investigators. et al. Effect of potentially modifiable risk factors associated with myocardial infarction in 52 countries (the INTERHEART study): case-control study. Lancet 2004;364:937–52.

[33] Iqbal R, Anand S, Ounpuu S, Islam S, Zhang X, Rangarajan S, INTERHEART Study Investigators. et al. Dietary patterns and the risk of acute myocardial infarction in 52 countries: results of the INTERHEART study. Circulation 2008;118:1929–37.

[34] Zhou BF, Stamler J, Dennis B, MoagStahlberg A, Okuda N, Robertson C, et al. Nutrient intakes of middle-aged men and women in China, Japan, United Kingdom, and United States in the late 1990s: the INTERMAP study. J Hum Hypertens 2003;17:623–30.

[35] Chauhan AK, Singh RB, Ozimek L, Basu TK. Saturated fatty acid and sugar; how much is too much for health? A scientific statement of the International College of Nutrition. View point. World Heart J 2016;8:71–8.

[36] Mozaffarian D, Ludwig DS. The 2015 US Dietary Guidelines lifting the ban on total dietary fat. JAMA 2015;313(24):2421–2.
[37] Singh RB, De Meester F, Wilczynszka A, Takahashi T, Juneja L, Watson RR. Can a changed food industry prevent cardiovascular diseases? World Heart J 2013;5:1–8.
[38] US Department of Health and Human Services; US Department of Agriculture. 2015–2020 Dietary Guidelines for Americans. 8th ed. Washington, DC: US Dept of Health and Human Services; December 2015. http://www.health.gov/DietaryGuidelines.
[39] Schwingshackl L, Hoffmann G. Diet quality as assessed by the healthy eating index, the alternate healthy eating index, the dietary approaches to stop hypertension score, and health outcomes: a systematic review and meta-analysis of cohort studies. J Acad Nutr Diet 2015;115(5):780–800.
[40] Wang DD, Li Y, Chiuve SE, Hu FB, Willett WC. Improvements in US diet helped reduce disease burden and lower premature deaths, 1999–2012; overall diet remains poor. Health Affairs 2015;34(11):1916–22.
[41] Hu EA, Pan A, Malik V, Sun Q. White rice consumption and risk of type 2 diabetes: meta-analysis and systematic review. BMJ 2012;344:e1454.
[42] Liang W, Lee AH, Binns CW. White rice-based food consumption and ischemic stroke risk: a case-control study in southern China. J Stroke Cerebrovasc Dis 2010;19:480–4.
[43] Gadgil MD, Appel LJ, Yeung E, et al. The effects of carbohydrate, unsaturated fat, and protein intake on measures of insulin sensitivity: results from the OmniHeart trial. Diabetes Care 2013;36:1132–7.
[44] Appel LJ, Sacks FM, Carey VJ, et al. Effects of protein, monounsaturated fat, and carbohydrate intake on blood pressure and serum lipids: results of the OmniHeart randomized trial. JAMA 2005;294:2455–64.
[45] Lagiou P, Sandin S, Lof M, et al. Low carbohydrate-high protein diet and incidence of cardiovascular diseases in Swedish women: prospective cohort study. BMJ 2012;344:e4026.
[46] Sieri S, Krogh V, Berrino F, et al. Dietary glycemic load and index and risk of coronary heart disease in a large Italian cohort: the EPICOR study. Arch Intern Med 2010;170:640–7.
[47] Kokubo Y, Iso H, Saito I, Yamagishi K, Ishihara J, Inoue M, et al. Dietary fiber intake and risk of cardiovascular disease in the Japanese population: the Japan Public Health Center-based study cohort. Eur J Clin Nutr 2011;65:1233–41.
[48] Eshak ES, Iso H, Date C, Kikuchi S, Watanabe Y, Wada Y, et al. Dietary fiber intake is associated with reduced risk of mortality from cardiovascular disease among Japanese men and women. J Nutr 2010;140:1445–53.
[49] Fan J, Song Y, Wang Y, Hui R, Zhang W. Dietary glycemic index, glycemic load, and risk of coronary heart disease, stroke, and stroke mortality: a systematic review with meta-analysis. PLoS ONE 2012;7:e52182.
[50] Steffen LM, Jacobs DR Jr, Stevens J, et al. Associations of whole-grain, refined-grain, and fruit and vegetable consumption with risks of all-cause mortality and incident coronary artery disease and ischemic stroke: the Atherosclerosis Risk in Communities (ARIC) Study. Am J Clin Nutr 2003;78:383–90.
[51] Malik VS, Popkin BM, Bray GA, et al. Sugar-sweetened beverages and risk of metabolic syndrome and type 2 diabetes: a meta-analysis. Diabetes Care 2010;33:2477–83.
[52] Bernstein AM, Rosner BA, Willett WC. Cereal fiber and coronary heart disease: a comparison of modeling approaches for repeated dietary measurements, intermediate outcomes, and long follow-up. Eur J Epidemiol 2011;26:877–86.
[53] Kataoka M, Venn BJ, Williams SM, Te Morenga LA, Heemels IM, Mann JI. Glycaemic responses to glucose and rice in people of Chinese and European ethnicity. Diabet Med 2013;30:e101–7.

[54] Nilsson LM, Winkvist A, Eliasson M, et al. Low-carbohydrate, high-protein score and mortality in a northern Swedish population-based cohort. Eur J Clin Nutr 2012;66:694–700.
[55] Sjögren P, Becker W, Warensjo E. Mediterranean and carbohydrate-restricted diets and mortality among elderly men: a cohort study in Sweden. Am J Clin Nutr 2010;92:967–74.
[56] Trichopoulou A, Psaltopoulou T, Orfanos P, Hsieh CC, Trichopoulos D. Low-carbohydrate-high-protein diet and long-term survival in a general population cohort. Eur J Clin Nutr 2007;61:575–81.
[57] Helnæs A, Kyrø C, Andersen I, Lacoppidan S, Overvad K, Christensen J, et al. Intake of whole grains is associated with lower risk of myocardial infarction: the Danish Diet, Cancer and Health Cohort. Am J Clin Nutr 2016;103(4):999–1007.
[58] Park Y, Subar AF, Hollenbeck A, Schatzkin A. Dietary fiber intake and mortality in the NIH-AARP diet and health study. Arch Intern Med 2011;171:1061–8.
[59] Johnsen NF, Frederiksen K, Christensen J, Skeie G, Lund E, Landberg R, et al. Whole-grain products and whole-grain types are associated with lower all-cause and cause-specific mortality in the Scandinavian HELGA cohort. Br J Nutr 2015;114(4):608–23.
[60] Martínez-González MA, Fernández-Jarne E, Losa EM, Prad Santamaría M, Brugarolas-Brufau C, Serrano-Martinez M. Role of fibre and fruit in the Mediterranean diet to protect against myocardial infarction: a case−control study in Spain. Euro J Clin Nutr 2002;56:715–22.
[61] Kelishadi R, Mansourian M, Heidari-Beni M. Association of fructose consumption and components of metabolic syndrome in human studies: a systematic review and meta-analysis. Nutrition 2014 May;30(5):503–10.
[62] Chiavaroli L, de Souza RJ, Ha V, Cozma AI, Mirrahimi A, Wang DD, et al. Effect of fructose on established lipid targets: a systematic review and meta-analysis of controlled feeding trials. J Am Heart Assoc 2015;4(9):e001700.
[63] Turati F, Dilis V, Rossi M, Lagiou P, Benetou V, Katsoulis M, et al. Glycemic load and coronary heart disease in a Mediterranean population: the EPIC Greek cohort study. Nutr Metab Cardiovasc Dis 2015;25(3):336–42.
[64] Bo Simona, Ponzo Valentina, Goitre Ilaria, Fadda Maurizio, Pezzana Andrea, Beccuti Guglielmo, et al. Predictive role of the Mediterranean diet on mortality in individuals at low cardiovascular risk: a 12-year follow-up population-based cohort study. J Transl Med 2016;14:91.
[65] Shin JY, Kim JY, Kang HT, Han KH, Shim JY. Effect of fruits and vegetables on metabolic syndrome: a systematic review and meta-analysis of randomized controlled trials. Int J Food Sci Nutr 2015;66(4):416–25.
[66] Singh RB, Rastogi SS, Singh R, Ghosh S, Niaz MA. Effects of guava intake on serum total and high-density lipoprotein cholesterol levels and on systemic blood pressure. Am J Cardiol 1992;70(15):1287–91.
[67] Singh RB, Rastogi SS, Singh NK, Ghosh S, Gupta S, Niaz MA. Can guava fruit intake decrease blood pressure and blood lipids? J Hum Hypertens 1993;7(1):33–8.
[68] Singh RB, Rastogi SS, Niaz MA, Ghosh S, Singh R, Gupta S. Effect of fat-modified and fruit- and vegetable-enriched diets on blood lipids in the Indian Diet Heart Study. Am J Cardiol 1992;70(9):869–74.
[69] Estruch R, Ros E, Salas-Salvadó J, Covas MI, Corella D, Arós F, PREDIMED Study Investigators. et al. Primary prevention of cardiovascular disease with a Mediterranean diet. N Engl J Med 2013;368(14):1279–90.
[70] Toledo E, Salas-Salvadó J, Donat-Vargas C, Pilar Buil-Cosiales P, et al. Mediterranean diet and invasive breast cancer risk among women at high cardiovascular risk in the PREDIMED trial: a randomized clinical trial. JAMA Intern Med 2015;5(11):1752–60.

[71] Salas-Salvadó J, Bulló M, Babio N, for the PREDIMED Study Investigators. et al. Reduction in the incidence of type 2 diabetes with the Mediterranean diet. Diabetes Care 2011;34:14–9.

[72] Singh RB, Hristova K, Fedacko J, Singhal S, Khan S, Wilson DW, et al. Antioxidant vitamins and oxidative stress in chronic heart failure. World Heart J 2015;7:257–64.

[73] Rajoria A, Kumar J, Chauhan AK. Anti-oxidative and anti-carcinoginic role of lycopene in human health-a review. J Dairy Foods Home Sci 2010;29:3–4.

[74] Ferre GM, Guasch-Ferré M, Bulló M, Martínez-González MÁ, Ros E, Corella D, et al. Frequency of nut consumption and mortality risk in the PREDIMED nutrition intervention trial. BMC Med 2013;11:164.

[75] Du H, Li L, Bennett D, Guo Y, for the China Kadoorie Biobank Study. et al. Fresh fruit consumption and major cardiovascular disease in China. N Engl J Med 2016;374:1332–43.

[76] de Lorgeril M, Renaud S, Mamelle N, Salen P, Martin JL, Monjaud I, et al. Mediterranean alpha-linolenic acid-rich diet in secondary prevention of coronary heart disease. Lancet 1994;343(8911):1454–9. Erratum in: Lancet 1995;345(8951):738.

[77] Singh RB, Dubnov G, Niaz MA, Ghosh S, Singh R, Rastogi SS, et al. Effect of an Indo-Mediterranean diet on progression of coronary disease in high risk patients: a randomized single blind trial. Lancet 2002;360:1455–61.

[78] Singh RB, Rastogi SS, Verma R, Bolaki L, Singh R, Ghosh S. An Indian experiment with nutritional modulation in acute myocardial infarction. Am J Cardiol 1992;69:879–85.

[79] Singh RB, Fedacko J, Vargova V, Niaz MA, Rastogi SS, Ghosh S. Effect of low W-6/W-3 ratio fatty acid Paleolithic style diet in patients with acute coronary syndromes. A randomized, single blind, controlled trial. World Heart J 2012;3:71–84.

[80] Fitó M, Guxens M, Corella D, Sáez G, Estruch R, de la Torre R, et al. Effect of a traditional Mediterranean diet on lipoprotein oxidation. Arch Intern Med 2007;167(11):1195–203.

[81] Siti HN, Kamisah Y, Kamsiah J. The role of oxidative stress, antioxidants and vascular inflammation in cardiovascular disease (a review). Vascul Pharmacol 2015;71:40–56.

[82] Guralp O. Effects of vitamin E on bone remodeling in perimenopausal women: mini review. Maturitas 2014;79(4):476–80.

[83] Lasa A, Miranda J, Bulló M, Casas R, Salas-Salvadó J, Larretxi I, et al. Comparative effect of two Mediterranean diets versus a low-fat diet on glycaemic control in individuals with type 2 diabetes. Eur J Clin Nutr 2014;68(7):767–72.

[84] Ajala O, English P, Pinkney. Systematic review and meta-analysis of different dietary approaches to the management of type 2 diabetes. Am J Clin Nutr 2013;97(3):505–16.

[85] Carter P, Achana F, Troughton J, Gray LJ, Khunti K, Davies MJ. A Mediterranean diet improves HbA1c but not fasting blood glucose compared to alternative dietary strategies: a network meta-analysis. J Hum Nutr Diet 2014;27(3):280–97.

[86] Aune D, Keum N, Giovannucci E, et al. Whole grain consumption and risk of cardiovascular disease, cancer, and all cause and cause specific mortality: systematic review and dose-response meta-analysis of prospective studies. BMJ 2016;353:i27161.

[87] Singh RB, Sharma JP, Rastogi V, et al. Social class and coronary artery disease in a rural population of North India. The Indian Social Class and Heart Survey. Eur Heart J 1997;18:1728–35.

[88] Singh RB, Verma SP, Niaz MA. Social class and coronary artery disease in India. Lancet 1999;353:154–5.

[89] Singh RB, Rastogi SS, Rastogi V, et al. Blood pressure trends, plasma insulin levels and risk factors, in rural and urban elderly populations of north India. Coron Artery Dis 1997;8:463–8.

[90] Janus ED, Postiglione A, Singh RB, Lewis B. The modernization of Asia: implications for coronary heart disease. Council on arteriosclerosis of the International Society and Federation of Cardiology. Circulation 1996;94:2671–3.

[91] Indian Consensus Group. Indian consensus for prevention of hypertension and coronary artery diseases: a joint scientific statement of Indian Society of Hypertension and International College of Nutrition. J Nutr Environ Med 1996;6:309–18.
[92] Singh RB, Hideki Mori, Junshi Chen, et al. Recommendations for the prevention of coronary artery disease in Asians: a scientific statement of the International College of Nutrition. J Cardiovasc Risk 1996;3:489–94.
[93] Singh RB, Rastogi SS, Rao PV, et al. Diet and lifestyle guidelines and desirable levels of risk factors for the prevention of diabetes and its risk factors in Indians: a scientific statement of the International College of Nutrition. J Cardiovasc Risk 1997;4:201–8.
[94] Singh RB, Niaz MA, Ghosh S, et al. Association of trans fatty acids (vegetable ghee), and clarified butter (Indian ghee) intake with higher risk of coronary artery disease,in rural and urban populations with low fat consumption. Int J Cardiol 1996;56:289–98.
[95] Singh RB, Otsuka K, Chiang CE, Joshi SR. Nutritional predictors and modulators of meta-bolic syndrome. J Nutr Environ Med 2004;14:3–16.
[96] Mensah GA, Brown DW. An overview of cardiovascular disease burden in the United States. Health Affairs 2007;26(1):38–48.
[97] Rayner M, Allender S, Scarborough P. Group BHFHPR. Cardiovascular disease in Europe. Eur J Cardiovasc Prev Rehab 2009;16(2 Suppl.):S43–7.
[98] Pestana JA, Steyn K, Leiman A, Hartzenberg GM. The direct and indirect costs of cardiovascular disease in South Africa in 1991. S Afr Med J 1996;86(6):679–84.
[99] Azambuja MI, Foppa M, Maranhao MF, Achutti AC. Economic burden of severe cardiovascular diseases in Brazil: an estimate based on secondary data. Arquivos brasileiros de cardiologia 2008;91(3):148–55. 163–71.
[100] Dalziel K, Segal L, De Lorgeril M. A mediterranean diet is cost-effective in patients with previous myocardial infarction. J Nutr 2006;136(7):1879–85.
[101] Gaspoz J-M, Coxson PG, Goldman PA, et al. Cost effectiveness of aspirin, clopidogrel, or both for secondary prevention of coronary heart disease. New Engl J Med 2002;346(23):1800–6.
[102] Abdullah MM, Gyles CL, Marinangeli CP, Carlberg JG, Jones PJ. Cost-of-illness analysis reveals potential healthcare savings with reductions in type 2 diabetes and cardiovascular disease following recommended intakes of dietary fiber in Canada. Front Pharmacol 2015;6:167.
[103] Zhou J, Martin RJ, Tulley RT, et al. Dietary resistant starch upregulates total GLP-1 and PYY in a sustained day-long manner through fermentation in rodents. Am J Physiol Endocrinol Metab 2008;295(5):E1160–6.
[104] Keenan MJ, Zhou J, Hegsted M, Pelkman C, Durham HA, Coulon DB, et al. Role of resistant starch in improving gut health, adiposity, and insulin resistance. Adv Nutr 2015;6(2):198–205.

FURTHER READING

DeSalvo KB, Olson R, Casavale KO. Dietary Guidelines for Americans. JAMA. 2016.
Post RE, Mainous AG 3rd, King DE, Simpson KN. Dietary fiber for the treatment of type 2 diabetes mellitus: a meta-analysis. J Am Board Fam Med 2012;25(1):16–23.

CHAPTER 7

Sources of Fiber

Rosa M. Esteban, Esperanza Mollá, Vanesa Benítez

1 INTRODUCTION

Dietary fiber (DF) is an important component of a healthy diet, which is naturally present in plant foods, such as cereals, vegetables, fruits, and nuts. The content, structure, chemical composition, and physico-chemical properties of DF, as well as its physiological effects in the organism, are different depending on the source of fiber [1].

DF has been defined and classified in different ways over the years [2]. There are definitions based on analytical methods, on physiological effects, as well as on solubility and fermentability/viscosity of the fiber components. The type and amount of fiber consumed have shown to play an important role in the prevention of different diseases.

According to the Codex Committee on Nutrition and Foods for Special Dietary Used [3], DF is defined as carbohydrate polymers with more than nine monomers, which are not digestible by the endogenous enzymes in the small intestine. DF includes: edible carbohydrate polymers, associated with lignin and/or other compounds, carbohydrate polymers that have been obtained from food raw material by physical, enzymatic, or chemical means, and which have been shown to have a physiological effect of benefit to health as demonstrated by generally accepted scientific evidence, and synthetic carbohydrate polymers that have been shown to have a physiological effect of benefit to health [4–6].

DF can be classified according to its solubility in water as soluble dietary fiber (SDF) and insoluble dietary fiber (IDF) [7,8]. Lignin, cellulose, and some hemicelluloses typically make up the bulk of IDF, whereas pectic polysaccharides, beta-glucans, galactomannans, fructans, gums, and other nonstarch polysaccharides make up SDF [6]. DF has shown protective effects toward cardiovascular disease, colorectal cancer, obesity, and diabetes [9]. Nevertheless, soluble and insoluble fractions of DF have different physiological effects. Thus, SDF is mainly related to blood cholesterol

Dietary Fiber for the Prevention of Cardiovascular Disease
http://dx.doi.org/10.1016/B978-0-12-805130-6.00007-0

lowering and to the decrease of glucose absorption in the small intestine, and some of DF components also have a potential prebiotic effect (selective stimulation of potential beneficial bacteria of the fecal microbiota following the consumption of foods containing those components). On the other hand, IDF is responsible for increasing fecal bulk, intestinal regulation, and water absorption [10].

SDF (pectic polysaccharides and other hydrocolloids) is mainly found naturally in fruits, vegetables, and legumes, while IDF, including cellulose and some hemicelluloses, is mainly found in whole grains [11]. The ratio of soluble to insoluble fiber, particle size and source of fiber are some of the important attributes for both functional and dietary properties. A ratio of approximately 1:2 or 1:3 of soluble to insoluble fiber is considered acceptable to use fiber as a food ingredient [12].

Physicochemical properties of DF are related to its functions and activities in the organism. Hydration properties partly determine the behavior of DF in the digestive tract and account for some of their physiological effects (fecal bulking of minimally fermented DF). DF viscosity can impede the digestion and absorption of nutrients from the gut. Moreover, DF can bind certain substances as ions or organic molecules, such as bile acids, cholesterol, or glucose, which can reduce their absorption [8].

Although the recommended daily intake of DF varies across countries, it is generally considered that an ideal fiber intake ranges between 20 and 35 g/day. However, few countries report fiber intakes that follow this recommendation; in fact, the DF consumption of most population in developed countries is about 11 g/day [6].

Nevertheless, in order to increase the DF daily intake of the population, manufacturers have developed in recent years fiber-enriched foods using high-fiber ingredients [13].

Traditionally, most of the DF intake comes from the cell walls of vegetables, fruits, cereal products, and other seeds. Vegetables, fruits, and cereals in the diet comprise a diversity of tissues. Among them, parenchymal tissues are particularly important in relation to DF, because the cell walls of these tissues are the main sources of DF from edible fruits and vegetables. Only in cereal bran products, lignified tissues are of great importance and have a significant contribution to DF [1].

In this chapter, different sources of DF have been reviewed and evaluated for their fiber content and composition of cereals, legumes, fruits, vegetables, and nuts. Moreover, we have also included a review of the main changes to DF as a consequence in processing of foods.

2 CEREALS AS SOURCE OF FIBER

Among fiber sources, cereal grains and their products are considered to be a good source of DF. According to Lambo et al. [14], cereals contribute to about 50% of fiber intake in western countries, and therefore they show a high contribution to the intake of nonstarch polysaccharides (the main components of DF).

Nevertheless, the content and composition of DF in cereal products vary with the cereal species, type of processing, and the extent of refinement of the grains; for instance, on milling of cereal grains, the removal of the bran fraction (aleurone and outer layers of grains) results in an important loss of DF [1,15].

2.1 Polysaccharide Composition of Dietary Fiber in Cereal Grains

DF of cereals is mainly composed of hemicelluloses, with smaller amounts of cellulose, pectin substances, and glycoproteins. In the case of wheat, the cell walls of endosperm (inner part of the grain) are rich in hemicelluloses consisting of highly branched arabinoxylans associated with phenolics [16], while rice endosperm cell walls contain cellulose in high amount, and also pectin polysaccharides and arabinoxylans [1]. Arabinoxylans act as prebiotic components and furthermore they enhance postprandial glucose and insulin sensitivity [17,18].

On the contrary, barley and oats hemicelluloses comprise mainly beta-glucans. Beta-glucans found in cereals are homopolysaccharides composed of glucose monomers with mixed linkages and they vary in solubility [19]. Oats and barley have a higher proportion of soluble beta-glucans than wheat [15]. Beta-glucans and other types of viscous soluble fibers reduce postprandial glucose and insulin responses after a meal and improve insulin sensitivity both in diabetic and nondiabetic individuals. In this sense, DF viscosity could be considered as a potential mechanism to reduce starch digestion rate and sugar absorption, inhibiting mixing and diffusion in the intestinal tract and delaying gastric emptying [15].

Studies comparing the efficacy of beta-glucans from different sources on lowering cholesterol levels have been made. Delaney et al. [20] observed no significant differences in cholesterol level reduction of hamsters fed with beta-glucans from barley and from oats. Similar conclusions were obtained from Hallfrisch et al. [21] in relation to the effects of beta-glucans from barley or oats on glucose and insulin response in humans.

2.2 Dietary Fiber Content in Cereals

Table 7.1 summarizes the DF content of raw cereals [9,22–25] and pseudo-cereals [22,26]. DF ranged from 1.1% to 17.3%, depending on the species. Barley and rye showed the highest content of DF, followed by oats, wheat, corn, and millet. *Sorghum*, rice, and fonio (an African cereal crop) showed lower content of DF. Nevertheless, the DF content of each cereal found in the literature is very heterogeneous, probably due to the various cereal cultivars.

According to Li et al. [25], cereals had total dietary fiber (TDF) content higher than 1 g/100 g, so that they might be considered good sources of DF, with different proportions of soluble and insoluble fractions. In all cases, the content of IDF was higher than that of SDF. SDF ranged from 0.3 g/100 g in rice to 3.8 g/100 g in oats. As it could be expected, the contents of DF in the whole grains (including bran) were higher than in the flour from endosperm (inner part of cereal grain), since bran is the richest source of DF of cereal grains (42.8 g/100 g of DF) in wheat bran. Wheat bran, as other cereal brans, is added to foods to increase the fiber content.

Barikmo et al. [24], in a study carried out on cereals from Mali, indicated that millet showed a mean value of fiber of 6.2 g/100 g, sorghum, and corn (maize) 4.7 and 4.6 g/100 g, respectively, while the fiber content of wheat had a mean value of 3.0 g/100 g. The mean values for fiber in fonio and rice were 2.2 and 1.1 g/100 g, respectively.

Rice (*Oryza sativa* L.) is an important cereal consumed by about half of the world's population and represents a major source of daily energy and nutrients. Rice provides 23% of global human energy and 16% of per capita protein [27]. Fiber content in rice depends on rice variety and the type of processing. Sumczymski et al. [28], in a study about IDF (cellulose, insoluble hemicelluloses, and lignin) in rice types commercialized in Czech markets, reported IDF values ranged between 0.69 and 8.22 g/100 g. The highest amount was provided by red rice and jasmine rice (brown and red), while the lowest IDF contents were found in parboiled rice, sushi rice, and basmati rice, all of them with values lower than 1%.

Cereal grains are good sources of beta-glucans, a cell wall carbohydrate, which has beneficial effects on human health, such as cholesterol-lowering properties [29]. Demirbas [23] studied the content of beta-glucans in different cereals and found that oat and barley grains had the highest beta-glucan contents, while rice, rye, and wheat had lower amounts. However, barley is mainly used for animal feed, as well as for malting and brewing in beer and whisky manufacture, whereas it is underutilized to obtain food ingredients [30].

Table 7.1 Dietary fiber content in raw cereals and pseudocereals

Cereals (g/100 g edible portion)	Scientific name	TDF	IDF	SDF	Beta-glucans	References
Barley	*Hordeum vulgaris*	14.8–17.3				[8,22]
Barley					2.8–7.2	[23]
Barley flour		10.1				[22]
Corn	*Zea mays*	13.4			0.5–1.3	[8,23]
Corn flour		4.6–9.4				[22,24]
Corn meal, degermed		3.9	3.3	0.6		[25]
Fonio	*Digitalia exilis*	2.2				[24]
Oats (raw)	*Avena sativa*	10.3–10.6	6.5	3.8		[8,22]
Oat flour		10.0			2.2–6.6	[22,23]
Millet	*Panicum miliaceum*	8.5			0.5–1.0	[22,23]
Millet flour	*Pennisetum glaucum*	6.2				[24]
Rice	*O. sativa*	1.1–1.3	1.0	0.3	0.4–0.9	[8,23,24]
Rice (brown)		3.0				[22]
Rye	*Secale cereale*	14.6			0.7–2.9	[22,23]
Rye flour		11.7				[22]
Sorghum (flour)	*Sorghum bicolor*	4.7				[24]
Wheat (ground)	*Triticum aestivum*	3.0–4.0				[22,24]
Wheat (whole grain)		10.3–12.6	10.2	2.3		[8,22]
Wheat flour (whole grain)		9.0				[22]
Wheat bran		42.8				[22]
Wheat germ		14.0–15.9	12.9	1.1		[8,22]
Pseudocereals						
Amaranth (defatted flour)	*Amaranthus caudatus*, L.	11.0	8.6	2.4		[26]
Quinoa (defatted flour)	*Chenopodium quinoa* W.	10.0	7.8	2.2		[26]
Quinoa (raw)		7.9				[22]

IDF, Insoluble dietary fiber; SDF, soluble dietary fiber; TDF, total dietary fiber.

Several studies have indicated that beta-glucans from both barley and oat have the effect of flattening the postprandial blood glucose and rising inulin, also reducing serum cholesterol levels [20,21].

The pseudocereals quinoa and amaranth have DF contents similar to those found in cereal grains. Lamothe et al. [26] analyzed the content and composition of DF of amaranth and quinoa seeds from Peru and Bolivia, respectively. They found that 78% of fiber was insoluble in both pseudocereals and was mainly composed of pectic polysaccharides (homogalacturonans and rhamnogalacturonans-I with arabinoxylan side-chains), as well as hemicelluloses (highly branched xyloglucans) and cellulose. The proportion of SDF was higher than in other cereals, such as wheat and maize. Cordeiro et al. [31] reported that pectic polysaccharides from quinoa seeds, consisting of linear arabinans and rhamnogalacturonan-I with arabinose and galactose side-chains, showed to have gastroprotective activity against ethanol-induced lesions in rats.

2.3 Dietary Fiber Content and Composition of Cereals as Affected by Processing

Cereals are processed in different ways before consumption. For this reason DF content, composition, and structural characteristics could be altered, and therefore the physico-chemical properties and physiological effects of fiber fractions might change.

Milling of cereals causes partial disruption of DF structures from endosperm, making the endosperm more accessible to digestive enzymes and DF more accessible to bacteria in the colon, thus affecting the extent of fermentation of DF components, and subsequently the physiological effects of DF. In this sense, particle size has a great importance in relation to physiological properties of DF, since excessive milling may have negative effects, such as a constipating effect, and will not exhibit the transit time normalization, associated to increased particle size [2].

Heat treatments may cause degradation of fiber matrix with a breakdown of glycosidic linkages of DF components, such as pectic polysaccharides, which can lead to cell separation and tissue softening [1]. This degradation might result in a solubilization of fiber, with redistribution between insoluble and soluble fractions, reducing in some cases TDF content. Changes in polysaccharide chain length due to heating may reduce physico-chemical properties, such as viscosity and water holding capacity of DF [2].

Dhingra et al. [8], in a review about DF in foods, concluded that food processing (mechanical, thermal, and thermo-mechanical) might affect DF content and composition, as well as its physico-chemical and functional properties.

Tables 7.2–7.4 shows the DF contents of various cereal products [22,25,32]. Of the 62 cereal based foods shown in the above-mentioned tables, only two products have less than 1 g/100 g of TDF, so that the remaining foods may be considered good sources of DF, according to Li et al. [25], with different proportions of soluble and insoluble fractions.

Table 7.2 Dietary fiber content in baked products

Baked products (g/100 g edible portion)	TDF	IDF	SDF	Beta-glucans	References
Breads					
Barley bread	4.6				[22]
Corn bread	2.4				[22]
Oat bread	4.0				[22]
Rye bread	6.5				[22]
Rye bread, seedless	4.5	2.8	1.6		[25]
Wheat bread, soft	3.4–6.2	2.1–4.6	1.3–1.6		[25]
White wheat bread	1.5–4.3[a]	0.5–1.4	1.0–1.3	0.2	[22,25,32]
Whole wheat bread	6.0–7.0[a]	4.8–5.2	1.3–1.5	0.3	[22,25,32]
Wheat/corn bread	3.4[a]			0.2	[32]
Wheat/rye bread	6.2[a]			0.4	[32]
Wheat bread (sliced, toasted)	3.6-4.5				[22]
Whole wheat bread toasts	4.5				[22]
White bread (reduced calorie)	9.5	8.5	1.0		[25]
Other baked products					
Bagel (plain, frozen)	2.5	1.3	1.2		[25]
Bakery	2.3				[22]
Butter cookies	1.6				[22]
Cookies	3.2				[22]
Cookies (“digestive” type)	4.6				[22]
Crackers (whole meal)	12.5				[22]
Croissants (chocolate)	3.0				[22]
Doughnuts	3.0				[22]
Hamburger/hotdog rolls	2.0	1.4	0.6		[25]
Hamburger type	4.1				[22]
Muffin	1.0				[22]
Tortilla (corn flour)	5.5	4.4	1.1		[25]
Tortilla (wheat flour)	2.4	0.8	1.5		[25]

IDF, Insoluble dietary fiber; SDF, soluble dietary fiber; TDF, total dietary fiber.
[a]Data are expressed as g/100 g dry matter.

Table 7.3 Dietary fiber content in breakfast cereals

Breakfast cereals (g/100 g edible portion)	TDF	Beta-glucans	References
Barley flakes	10.8[a]	4.8	[32]
Corn flakes	2.9[a]	0.7	[32]
Corn (with honey)	3.0		[22]
Oat flakes	15.9[a]	5.1	[32]
Rye flakes	15.9[a]	1.9	[32]
Rice (puffed, enriched)	1.4		[22]
Rice (with chocolate)	1.7		[22]
Wheat (with sugar)	3.5		[22]
Wheat (with chocolate)	4.5		[22]
Whole wheat (all bran type)	29.0		[22]
Corn/ wheat/oat	2.4		[22]
Rice /wheat (with fruits)	5.8		[22]
Wheat/corn	3.5		[22]
Wheat/rice	2.5		[22]
Wheat/oat/corn (with honey and nuts)	4.4		[22]
Muesli based cereals	6.1		[22]

TDF, Total dietary fiber.
[a]Data are expressed as g/100 g dry matter.

Breads (Table 7.2) showed DF contents comprised between 1.5 and 6 g/100 g, with the highest contents in rye bread and whole wheat bread. SDF in breads ranged from 1.0 to 1.6 g/100 g, whereas IDF ranged from 0.5 g/100 g in white wheat bread to 4.8 g/100 g in whole wheat bread.

Other baked products were also analyzed, such as muffins and butter cookies, which showed DF contents around 1 g/100 g, and whole meal crackers, which showed a high level of DF (12.5 g/100 g). In Table 7.3, DF contents of various breakfast cereals are shown. All bran type cereals contained the highest amount of TDF (29.0 g/100 g), followed by oat flakes and rye flakes (both with 15.9 g/100 g of DF), while corn flakes showed lower DF content (2.0–3.0 g/100 g). Furthermore, puffed rice showed the lowest content of TDF.

Table 7.4 shows DF contents of different cereal products, such as cooked cereals, pasta, and other products. Among cooked cereals, rice (white long grain) had the lowest content of DF, 0.3 g/100 g. Cooked pasta showed DF contents between 1.0 and 4.0 g/100 g, depending on whether it came from endosperm meal or from whole grain meal.

Extrusion cooking, an important process used in many cereals, may cause changes in DF content. Several reports [33,34] indicated that extrusion might cause an increase in total DF content, and a redistribution of

Table 7.4 Dietary fiber contents in cereal products

Cereal products (g/100 g edible portion)	TDF	IDF	SDF	Beta-glucans	References
Cooked cereals					
Corn (cooked)	8.1[a]			3.1	[32]
Millet (cooked)	5.0[a]			—	[32]
Rice (cooked)	0.7–2.2[a]	0.7	—	—	[8,22,32]
Rice (white, long grain, cooked)	0.3	0.3			[25]
Rice (brown, long grain, cooked)	3.3	2.9	0.4		[25]
Wheat (cooked)	12.9[a]			0.6	[32]
Pasta					
Pasta (raw)	5.0				[22]
Pasta (whole meal, raw)	11.5				[22]
Pasta with vegetables (raw)	6.0				[22]
Pasta (boiled)	1.0				[22]
Pasta (whole meal, boiled)	4.0				[22]
Spaghetti (cooked)	2.1	1.3	0.5		[25]
Other products					
Appetizers (wheat)	6.3				[22]
Chips (corn)	4.9				[22]
Crusts (wheat)	6.3				[22]
Grits, instant (cooked)	1.6	1.5	0.1		[25]
Oatmeal (boiled)	0.8				[22]
Oatmeal, instant (cooked)	2.6	1.1	1.4		[25]
Pop corn (without oil and salt	15.0				[22]
Soluble cereals (powder)	2.3				[22]

IDF, Insoluble dietary fiber; SDF, soluble dietary fiber; TDF, total dietary fiber.
[a]Data are expressed as g/100 g dry matter.

insoluble to soluble fiber, while other authors [35] reported small decreases in total and insoluble fiber in extruded cereals, as a result of breaking of chemical bonds in polysaccharides, yielding more soluble fragments, and increasing soluble fiber fraction. Varo et al. [36], in a collaborative study of eight laboratories, indicated the effects of several thermal treatments in cereals and potatoes and reported no changes in DF content and starch as a result of extrusion. Wang et al. [35] studied the effects of extrusion on physico-chemical properties of fiber from cereals. These authors indicated a higher water absorption in extruded cereals, probably due to starch gelatinization, while Ralet et al. [37] reported no significant variation in water absorption capacity of extruded wheat bran.

Honcu et al. [38], in a study on extrusion effects on the content and properties of DF components in various barley cultivars, reported that the content of arabinoxylan was not affected significantly by extrusion, whereas beta-glucan and SDF content increased in some extruded barley cultivars. On the contrary, extrusion caused a significant decrease in IDF of all extruded barley flours.

Dodevska et al. [32] carried out a study on TDF components, including fructans (an indigestible carbohydrate with prebiotic effects) of various cereal products used in Serbian diet. The authors reported values of DF ranging from 2.5 to 20.9 g/100 g (on a dry matter basis). Flakes of oat and rye showed the highest contents of TDF, while corn flakes had low levels of DF (4.7 g/100 g). Cooked cereals had TDF contents ranging from 2.5 g/100 g in polished rice to 15.8 g/100 g in wheat, while whole meal rice showed intermediate levels (9.2 g/100 g). In this sense, processing of rice by abrasion or parboiling removed the seed coat and hence, a great part of DF was lost. The analysis of different breads showed TDF levels from 4.8 g/100 g in wheat/corn bread to 8.6 g/100 g in whole wheat bread. Regarding fructans, these compounds represented a remarkable proportion of TDF in some cereal products, with levels varying in wide range from 0.3 to 5.0 g/100 g. Among different cereal flakes, rye flakes had the highest content of fructans (5.0 g/100 g), whereas oat flakes the lowest one. Corn flakes, one of the most consumed breakfast cereals, showed intermediate levels of fructans (1.9 g/100 g). Among breads, wheat/rye bread was the richest source of fructans, while wheat/corn bread and white wheat bread had the lowest contents of these indigestible carbohydrates. Fructan content in cooked cereals varied from 0.3 g/100 g in polished rice to 2.9 g/100 g in wheat.

The composition of fiber was different depending on the type of cooked cereal and cereal product. In cooked wheat, the main fiber fraction was arabinoxylan, while in cooked millet, cooked sweet corn, and polished rice, resistant starch was the major fiber fraction [32]. Nevertheless, rice fructans was present in high amounts in whole meal, and cellulose was also a main component of TDF. These results showed that in rice grains, fructans, and other fiber components is mainly located in the bran. On the other hand, the occurrence of beta-glucans, an important fiber fraction with high health potential, was only detected in cooked wheat and whole meal rice [32]. However, Dermibas [23] reported the presence of low levels of beta-glucans in millet, corn, and rice from Turkey. Brennan and Cleary [30] indicated that the chemical structure, degree of polymerization, and molecular interactions of beta-glucans may be altered by processing of foods. These modifications may result in changes of physico-chemical properties and hence, affect the beneficial physiological effects of beta-glucans.

Dodevska et al. [32] studied the composition of TDF of different types of bread, showing that fructans, arabinoxylans, and resistant starch were the main components of fiber fractions in all breads analyzed, with minor amounts of beta-glucans and cellulose.

DF of oat and barley flakes was mainly constituted by beta-glucans with minor amounts of arabinoxylans. On the contrary, this latter compound was the major component of fiber in rye flakes. It is worth mentioning that corn flakes had the lowest amount of DF with resistant starch as the main component [32].

3 LEGUMES AS FIBER SOURCE

3.1 Dietary Fiber Content in Legumes

Legumes have been reported as good source of DF [39]. Table 7.5 summarizes the DF content of raw conventional legumes including chickpeas, beans, lentils, and peas. DF of conventional pulses ranged from 15.3% to 29.7%, depending on the species and varieties. In general, beans were the conventional legumes with the highest DF content, followed by chickpeas and peas, and the lowest contents were found in lentils, although the DF content of each pulse found in the literature was very heterogeneous, probably due to the great quantity of legume varieties existing in different countries. Bean DF content varied from 21.8% to 29.8 %, being white bean, kidney bean, and pinto bean the varieties with higher content [40–47]. Regarding lentils, their DF content ranged from 16.5% to 27.9%, standing out the high content reported in the variety Rubia de la Armuña [42,44,45,48,49]. DF content of peas is reported by, Mallillin et al. [40], Martín-Cabrejas et al. [50], and Almeida Costa et al. [44] in three cultivars, finding levels ranging between 15.3% and 29.7%. In relation to chickpea DF content (17.6%–27.2 %), Castellano chickpea has the lowest content [40,42,43,48,49].

The DF content of raw nonconventional pulses is compiled in Table 7.6. Nonconventional legumes include underexploited legumes in industrialized countries, such as soybean, cowpea, Jack bean, lupin, mucuna, or dolichos. These pulses are very interesting since they are used as potential sources of plant proteins in many developing countries [51,52]. Nonconventional legumes showed, in general, higher level of DF (20%–54.8%) than conventional ones. Soybean, mucuna, and dolichos stand out as legumes with high DF content, more than 40% [40,52,53]. Mungbean, pole sitao, cowpea, Jack bean (*Canavalia ensiformis*), and lupin presented also high level of DF, between 30% and 40% [40,47,52,53]. However, lima bean, pigeon pea, green gram, and some *Canavalia* spp. showed contents of DF slightly above 20%, similar or lower to those found in conventional pulses [40,49,54].

3.2 Legume Dietary Fiber Composition

DF of legumes, conventional, and underexploited ones, was mainly composed of insoluble fiber, which represented 62%–99% of TDF depending on species and varieties (Tables 7.5 and 7.6). The lower contributions of IDF were found in beans, since IDF represented between 66% and 80% of TDF in most varieties. In the rest of legumes, IDF fraction contributed more than 80% or 90% to the TDF.

Regarding SDF, the content of this fraction was different depending on the legume species. Pink mottle cream bean, pea cv. Esla, and lupin presented the largest contents of this fraction (25, 37, and 38%, respectively). The proportion of IDF and SDF is important for physiological and technological properties of DF, thus 30%–50% of SDF and 70%–50% of IDF is considered a well-balanced proportion to obtain the physiological effects of both fractions [55].

In general, IDF of legumes is mainly composed of hemicelluloses and cellulose, although this composition also varies with the species [56]. IDF of chickpea and lentil is composed mainly of arabinans and cellulose [48], while the main polysaccharides of IDF in peas are cellulose and pectic polysaccharides [50]. Regarding to nonconventional legumes, the bulk of IDF is mainly constituted by cellulose, resistant starch, hemicelluloses, and pectic polysaccharides [53].

In relation to SDF composition, pectin polysaccharides stand out as the main components of this fraction in beans, peas, and chickpeas. However, SDF fraction in lentil and cowpea contains low content of pectin polysaccharides. These low contents of pectin polysaccharides in SDF fraction were also found in other nonconventional legumes as dolichos, mucuna, and Jack bean [46,48,50,53].

3.3 Effects of Processing on Legume Dietary Fiber

Legume seeds are usually processed before being consumed in order to reduce the presence of antinutritional factors and to improve their sensorial characteristics. Dehydration, cooking, canned, germination, or fermentation are some of the processes to which legumes are subjected, and their effects on DF content and composition have been studied and reported.

Aguilera et al. [43] and Martin-Cabrejas et al. [46] studied the effect of dehydration process on DF of common legumes. This process involves

Table 7.5 Dietary fiber content in raw common legumes

Legumes (g/100 g edible portion)	Scientific name	TDF	IDF	SDF	References
Beans					
"Kidney"	*Phaseolus vulgaris* L.	29.8	29.4	0.4	[49]
"Pinto"	*P. vulgaris* L.	23.3	19.8	3.5	[41]
"White"	*P. vulgaris* L.	26.3–27	20.4–21.2	5.8–5.9	[42,43]
"IAC Carioca Ete"	*P. vulgaris* L.	22.3	19.9	2.4	[44]
"Tolosa"	*P. vulgaris* L.	27.3	21.5	5.8	[45]
"Carilla"	*P. vulgaris* L.	24.5	17.1	7.7	[46]
"Bean"	*P. vulgaris* L.	21.9	14.4	7.5	[47]
Pink mottle cream	*P. vulgaris* L.	21.8	16.4	5.4	[43]
Chickpea	*Cicer arietinum* L.	17.6–27.2	12.6–25.9	0.9–5	[40,42,43,48,49]
Lentils					
Lentil spp.	*Lens culinaris* L.	16.5–24.3	15.6–21.6	0.9–2.7	[42,48,49]
"Silvina"	*L. culinaris* L.	20.4	19	1.4	[44]
"Rubia de la Armuña"	*L. culinaris* L.	27.9	24.5	2.4	[45]
Peas					
Pea spp.	*Pisum sativum* L.	29.7	27.6	2.1	[40]
"Maria"	*P. sativum* L.	22.0	20.3	1.7	[44]
"Esla"	*P. sativum* L.	15.3	9.7	5.6	[50]

IDF, Insoluble dietary fiber; SDF, soluble dietary fiber; TDF, total dietary fiber.

Table 7.6 Dietary fiber content in raw nonconventional legumes

Legumes (g/100 g edible portion)	Scientific name	TDF	IDF	SDF	References
Cowpea	*Vigna unguiculata* (L.) Walp.	31.2–34.1	29.8–31.0	0.9–4.0	[40,52,53]
Dolichos	*Lablab purpureus* (L.) Sweet	42.0–48.0	39.9–44.0	2.1–4.0	[52,53]
Green gram	*Phaseolus aureus*	20.0	16.6	3.4	[49]
Jack bean	*Canavalia ensiformis* (L) DC	33.2–45.6	31.7–42.5	1.5–3.6	[52,53]
Lima bean	*Phaseolus lunatus*	20.9	17.7	3.7	[40]
Lupin	*Lupinus albus* cv. multolupa	38.0	23.6	14.4	[47]
Maunaloa	*Canavalia cathartica*	23.6	22.6	1.0	[54]
Mucuna	*Stizolobium niveum*	42.7–45.0	39.8/41.0	2.9/4.0	[52,53]
Mung bean	*Vigna radiate* (L.) Wilczek	31.7	26.9	4.8	[40]
Pigeon pea	*Cajanus cajan*	21.8	19.4	2.4	[40]
Pole sitao	*V. unguiculata* subsp. *Sesquipedalis* (L.) Verdc.	35.0	29.5	5.5	[40]
Soybean	*Glycine max*	46.9–54.8	38.9–52.1	2.7–8.0	[40,52]
Sword bean	*Canavalia gladiata*	21.0–22.0	20.0–21.0	0.02–0.8	[54]

IDF, Insoluble dietary fiber; SDF, soluble dietary fiber; TDF, total dietary fiber.

three consecutive steps: soaking, cooking, and dehydration. In general, the first step (soaking) did not change the TDF content of legumes. However, soaking plus cooking produced a significant increase of TDF (10%–17%), which was due to increases of SDF (33%–56%) in beans and chickpea, and increases of IDF in lentils and chickpea (31% or 19%, respectively). The dehydration stage showed further increases in both SDF and IDF fraction in the cooked legumes, contributing to TDF increments of 12%–27%. Therefore, the industrial dehydration process affected DF in legumes, resulted in significant increases of TDF.

However, Kutos et al. [41] found considerable decrease of IDF and TDF contents during cooking and canning of beans. Moreover, bean IDF content decreased during soaking, whereas SDF content increased. Cooking of soaked beans and canning of beans decreased the SDF content. Almeida-Costa et al. [44] studied the combined effect of the thermal treatment (cooking) together with freeze-drying in conventional legumes. In general, no significant differences were found between DF contents of raw and processed legumes. In the same way, Marconi et al. [57] demonstrated that cooking, both boiling, and microwave treatment, did not modify the total nonstarch polysaccharide content of beans and chickpeas compared to raw legumes, but both processes produced increases in SDF and decreases in IDF. These changes were attributed to a partial solubilization and depolymerization of hemicelluloses and insoluble pectin polysaccharides. Cooking and thermal processing resulted in a softening of plant tissues, due in part to the degradation of pectin polysaccharides that are involved in cell-to-cell adhesion in the middle lamella of the cell walls [58].

Aldwairji et al. [59] studied the fiber content of boiled and canned legumes commonly consumed in the United Kingdom. Results showed that the boiled legumes had TDF levels ranging from 3.6% in green beans to 11.2% in red kidney beans, whereas canned legumes showed a range of TDF values from 2.7% in canned green beans to 7.4% for canned chickpeas. Thus, legumes preserved by canning were found to have significantly lower TDF values than boiled ones. Moreover, canning did not significantly change the proportion of IDF to SDF compared to boiled legumes.

Germination is considered a suitable procedure to improve the nutritional value of legume seeds by reducing levels of antinutritional factors [47]. Several authors have reported the impact of germination on the DF contents in both conventional and nonconventional legumes. The results showed that the changes in DF content were dependant on the studied legume and germination conditions [47,49,50,52,53]. In general, germination process increased both IDF and SDF fractions, producing an enhancement on TDF content. The changes promoted by germination resulted in an improvement in IDF/SDF ratio. This fact might exert an influence on the physiological effects and on physico-chemical properties of DF [49,50,52,53,60]. Regarding the effect of fermentation on DF of beans, it was reported that both natural and microorganism fermentation produced slight decreases of TDF [46].

4 FRUITS, VEGETABLES, AND OTHER FIBER SOURCES

Fruits, vegetables, and nuts have been reported as DF sources whose composition and amount vary from one plant food to another. The contributions of fruits and vegetables to fiber intake in western countries are about 30%–40% for vegetables, and 16% for fruits [14].

The main constituents of DF from fruits and vegetables are hemicelluloses, pectic polysaccharides, and cellulose. Hemicelluloses are largely composed of xylose and glucose in fruits and vegetables, xyloglucans and glucuronoxylans being the main constituents [2]. Glycoproteins are also present in small amounts in the cell walls of soft tissues [1], and lignins appear in lignified tissues of fruits and vegetables, such as the small seeds covering strawberries. Nevertheless, lignin represents only a small part of DF in most fruits and vegetables, although it can be as high as 4% of dry matter in mature pears [2].

4.1 Fruits

Table 7.7 summarizes the DF content of some fruits including, among others, berries (such as black currant, raspberry, or strawberry), drupes (such as pear, mango, apricot, apple, or peach), cucurbits (melon and watermelon), or tropical fruits (such as pineapple, banana, or papaya) [2,8,22,25,61,62]. DF contents of fruits ranged from around 0.3 g/100 g to 7 g/100 g of edible portion, depending on the species and varieties. Black currant and raspberry showed the highest content of DF, while watermelon the lowest one. Li et al. [25] reported contents of SDF ranging from 0.04 g/100 g in ripe pineapple to 1.5 g/100 g in guava, whereas IDF contents were between 0.3 and 11.8 g/100 g in watermelon and ripe guava, respectively.

Table 7.7 Dietary fiber content in fruits

Fruits (g/100 g edible portion)	TDF	IDF	SDF	References
Apple	1.0–2.3	1.5–1.8	0.2–0.7	[2,8,22,61,62]
Apricot	2.1			[22,61]
Avocado	1.8–6.7	5.5	1.2	[22,25,61,62]
Banana	1.0–3.4	1.2	0.5–0.6	[2,8,22,25,61,62]
Black currant	7			[22]
Figs	2.5			[22,62]
Grapefruit	0.9	0.3	0.6	[25]
Grapes (white)	0.7–2.2	0.3–0.7	0.5–0.6	[2,8,22,61,62]
Guava	12.7	11.8	1.5	[25]
Kiwi	1.6–3.4	2.6	0.8	[8,22,61,62]
Mango	1.8–2.3	1.1	0.7	[22,25]
Melon	0.6–1.0			[2,22,61,62]
Nectarine	2.0	1.1	1.0	[25]
Orange	1.0–2–2.4	0.7–1	1.1–1.4	[2,8,22,25,61,62]
Papaya	1.9			[22]
Peach	1.4–2.3	1.0–1.2	0.8–0.9	[2,8,22,25,61,62]
Pear	1.0–3.2	2.0–2.2	1	[2,8,22,25,61]
Persimmon	1.4–2.5			[22,62]
Pineapple	0.6–1.5	1.1–1.4	0.04–0.2	[2,8,22,25,61]
Pomegranate	0.6–3.5	0.5	0.1	[8,22]
Plum (yellow)	2.1–2.9	1.8	1.1	[22,25,61,62]
Raspberry	6.7			[22]
Strawberry	1.0–2.3	1.3–1.7	0.6–0.9	[2,8,22,61,62]
Tangerine	1.8–1.9	1.4	0.4	[22,61,62]
Watermelon	0.3–0.5	0.3	0.1–0.2	[8,22,25,61,62]

IDF, Insoluble dietary fiber; SDF, soluble dietary fiber; TDF, total dietary fiber.

4.2 Vegetables and Mushroom

Table 7.8 compiles DF content of some vegetables and mushrooms [2,8,22,25,62]. Among vegetables, artichoke and beetroot showed the highest contents of fiber, while truffles had the highest DF content among mushrooms. Crucifers, such as Brussels sprout, broccoli, various types of cabbages, and cauliflower showed TDF contents ranging from 1.8 to 4.2 g/100 g, with contents of SDF fraction around 0.3–0.5 g/100 g. Asparagus, celery, lettuce, pepper, potato, pumpkin, and tomato showed TDF contents lower than 2.0 g/100 g. In all vegetables and mushrooms reported, IDF constituted the main fraction of DF.

Table 7.8 Dietary fiber content in raw vegetables and mushrooms

Vegetables and mushrooms (g/100 g edible portion)	TDF	IDF	SDF	References
Artichoke	9.4			[22]
Asparagus	1.5			[8,62]
Asparagus (green)	1.7			[22]
Beans (green)	1.9	1.4	0.5	[8]
Beetroot	7.8	5.4	2.4	[8]
Broccoli	3.0–3.3	3.0	0.3	[8,22]
Brussels sprout	2.3–4.3			[2,22]
Cabbage (green)	2.2	1.8	0.5	[25]
Cabbage (red) Lombard	2.5			[22]
Cabbage (white)	4.2			[22]
Carrots	2.5–2.9	2.3–2.4	0.2–0.5	[8,25]
Cauliflower	1.8–2.6	1.1–2.2	0.5–0.7	[8,22,25,62]
Celery	1.5	1.0	0.5	[8]
Cucumber (peeled)	0.5–1.1	0.5–0.9	0.1–0.2	[2,8,22,25]
Eggplant	2.4–6.6	5.3	1.3	[8,22]
Leek	2.8			[22]
Lettuce	1.0–1.5	0.9	0.1	[22,25]
Mushroom	2.0–2.5			[2,22]
Onion	1.8–2.2	1.2–2.2	0.7	[8,22,25]
Pepper (sweet, green)	1.5	1.0	0.5	[25]
Potato (without skin)	1.3–2	1.0	0.3	[8,22,62]
Pumpkin	0.5–2.4			[22,62]
Spinach	2.6–6.3	2.1–2.4	0.5–0.8	[8,22,25,62]
Tomato	1.2–2.8	0.8–1.2	0.1–0.4	[8,22,25]
Truffle	16.5			[25]

IDF, Insoluble dietary fiber; SDF, soluble dietary fiber; TDF, total dietary fiber.

Li et al. [25] studied the content of DF in edible portions of raw and cooked vegetables. These authors reported contents of TDF in cooked vegetables ranging from 2.0 to 4.7 g/100 g in potatoes (boiled) and broccoli (microwaved), respectively.

4.3 Other Sources of Dietary Fiber

In addition to fruits and vegetables as fiber sources, there are various plant foods, other than cereals and legumes, which could contribute to DF intake. Such is the case of nuts, dried fruits, various seeds, and some common species used as condiments. The DF contents in nuts, various seeds, and dried fruits are compiled in Table 7.9 [8,22,25,62]. Flaxseeds showed the highest content of DF (22.3 g/100 g of edible portion), as opposed to sunflower seeds (2.7 g/100 g). Nuts showed contents of DF ranging around

Table 7.9 Dietary fiber content in nuts, dried fruits and various seeds

Nuts, dried fruits, seeds (g/100 g edible portion)	TDF	IDF	SDF	References
Almond	11.20–14.3	10.1	1.1	[8,22,62]
Coconut	9.0	8.5	0.5	[8,22]
Date	7.1–8.7			[62]
Flaxseed	22.3	10.1	12.2	[8]
Hazelnut	8.4–10.0			[22,62]
Peanut	8.1			[22]
Pine nut	8.0–8.5	7.5	0.5	[62]
Pistachio nut	6.5			[22]
Prunes	8.1	3.6	4.5	[25]
Raisin	3.1–6.5	2.2	0.9	[22,25]
Sesame seed	7.8	5.9	1.9	[8]
Sunflower seeds (peeled)	2.7			[62]
Walnut	5.2			[8,22]

IDF, Insoluble dietary fiber; SDF, soluble dietary fiber; TDF, total dietary fiber.

5.2 g/100 g in walnut and 14.3 g/100 g in almonds. The proportion of IDF was much higher than that of SDF. Dried fruits, such as seedless raisins or prunes showed TDF contents of 3.1 and 8.1 g/100 g, respectively [25]. Besides their high DF contents, these kind of fruits are very interesting for the occurrence of beneficial compounds, such as antioxidants and vitamins. However, it is noteworthy that nuts and dried fruits are consumed less frequently and in smaller amount than fresh fruits.

Table 7.10 summarizes DF content of some common spices and seasonings commonly used. Cinnamon powder, dried oregano, and dried basil showed DF contents higher than 40 g/100 g. On the contrary, saffron, fresh parsley and mint, and garlic powder showed DF content lower than 10 g/100 g. These species are commonly used and much appreciated for their sensory properties, such as flavor, color, or pungency. Spices have never been considered as a potential source of fiber due to the small amounts used, however their DF content were high, as shown in Table 7.10, and thus they should be taken into account.

5 AGROINDUSTRIAL BY-PRODUCTS AS HIGH-FIBER FOOD INGREDIENTS

The importance of DF for human health has led to the development of a large and potential market for fiber-rich products and ingredients and there is a trend to find new sources of DF that could be used as fiber-rich ingredients in the food industry [63,64].

Table 7.10 Dietary fiber content in common spices and seasonings

Spices and seasonings (g/100 g edible portion)	TDF	References
Basil (dried)	40.5	[22]
Bay leaf	26.3	[22]
Cinnamon (powder)	54.3	[22]
Cumin	10.5	[22]
Dill (dried)	37.6	[22]
Garlic (powder)	9.0	[22]
Mint (fresh)	6.8	[22]
Oregano (dried)	42.8	[22]
Nutmeg	20.8	[22]
Paprika (powder)	20.0	[22]
Parsley (fresh)	5.0	[22]
Pepper (black)	26.5	[22]
Pepper (white)	26.2	[22]
Rosemary (dried)	24.1	[22]
Saffron	3.9	[22]
Thyme (dried)	18.6	[22]

TDF, Total dietary fiber.

The most known and consumed DF enriched foods are breakfast cereals and bakery products as integral breads or cookies [13,65], as well as milk and meat derived products. In order to choose a fiber source suitable for these products it is important to consider the nutritional quality of the fiber source, its caloric value, antioxidant capacity, and its content of total and soluble fiber, as well as the extent of fermentability and water retention of that fiber source [10,64].

Fiber must fulfill some requirements to be acceptable as food ingredient. Among others, it must have a high concentration in small quantities, a balanced composition of soluble and insoluble fiber, a bland taste, a good shelf life, as well as be compatible with food processing. Moreover, it is necessary that fiber have an adequate amount of associated bioactive compounds, a reasonable cost, and a positive image for consumers. However, the fiber must not have antinutritional components, an offensive odor, or negative color and textural effects [12].

Traditionally the enrichment consisted in the addition of unrefined cereals or bran; however, industry is starting to use other DF sources, as fruit and vegetable pomaces, since they present better nutritional quality, higher amount of soluble fiber, stronger antioxidant capacity, and greater grade of fermentability and water retention than the traditional ones [66–68].

DF market is constantly looking for new sources of fibers with healthy properties to satisfy the growing consumer requests. For these purpose, there are a great variety of agroindustrial by-products available, which may be used as high DF ingredients, among other uses [69].

According to Brennan and Clearly [30], an important consideration is related to the inclusion of food systems in the appropriate levels of beta-glucans. These compounds can modify the structure and characteristics of foods, altering its viscosity and structure, although these changes could be involved in the beneficial properties of beta-glucan rich foods [30].

Beta-glucans have been used successfully in cereal-based foods, such as pasta and breads [70–72], as a functional ingredient with effects on the rate of starch degradation and subsequently a decrease in glycemic index.

Furthermore, the addition of beta-glucans can influence positively the processing of foods, as in the case of certain dairy products. Brennan et al. [73] used beta-glucans and other SDF in the manufacture of low-fat ice creams and yogurts, improving the rheological characteristics and sensory properties of these products. However, in a study on the inclusion of beta-glucans in cheese [74], an alteration of rheological properties was observed. Therefore, the effect of beta-glucans on rheological and sensory properties depends on the type of food to which beta-glucans are incorporated.

Many fruits, as orange, lemon, apple, peach, grapes, pineapple, pomegranate, or olive, are used to produce juices. The recovered residues (pomaces) of the juice extraction are rich in different high-added value compounds, as high DF residues [75–77]. There are also several vegetables, such as pepper, artichoke, onion, or asparagus, which originate by-products rich in DF and other value added compound during their processing [64]. In recent years, much research have been developed to characterize and evaluate the properties of by-products, as well as to know the potential applications of these by-products in order to assess their use as new high-fiber ingredients [75–80].

Lopez-Marcos et al. [80] evaluated and compared the characteristics (proximate analysis, technological, and physicochemical properties) of co-products from lemon, grapefruit, and pomegranate juice extraction; from lemon ice cream production; and from tiger nuts (horchata) making process as DF sources. The results indicated that extracts rich in DF obtained from these agroindustrial coproducts could be used as functional ingredients for their content both in SDF and IDF, as well as for their technological and physiological properties. Pomegranate DF stood out for its promising results mainly related to cholesterol adsorption capacity.

Amaya-Cruz et al. [81], looking for sources of antioxidant DF, characterized mango, peach, and guava by-products obtained from juice industry and evaluated their effect on obesity related to hyperglycemia and hepatic osteatosis. Mango and peach by-products represented a good source of antioxidant DF due to their adequate balance of SDF and IDF, and their high content of polyphenols and carotenoids, while guava by-products was rich in IDF.

In other recent study, some physical and chemical properties of watermelon rind and peel powders of sharlyn melon, as well as their utilization as a partial replacement of wheat flour in cake making has been evaluated [82]. Sharlyn melon peels had higher fiber content than watermelon rind although both products were good sources of DF and phenolic compounds. Watermelon rind and sharlyn melon peel were good in antioxidant capacity to increase shelf life of cake. It was recommended substitution of wheat flour at 5% to produce an acceptable cake with no significantly different sensory properties than control.

REFERENCES

[1] Esteban RM, Mollá E, Valiente C, Jaime L, López-Andréu FJ, Martín-Cabrejas MA. Dietary fibre: chemical and physiological aspects. Recent Res Dev Agric Food Chem 1998;2:293–308.
[2] Mongeau R, Brooks SPJ. Dietary fiber: properties and sources. Encyclopedia Food Health 2016;2:404–12.
[3] Codex Alimentarius Commission. Thirty second Session: Report of the 30th Session of the Codex Committee on Nutrition and Foods for Special Dietary Uses. Cape Town, South Africa, 3–7 November, 2008. Rome; 2009.
[4] Li C, Uppal M. Canadian Diabetes Association National Nutrition Committee Clinical Update on dietary fibre in diabetes: food sources to physiological effects. Can J Diab 2010;34(4):355–61.
[5] Almeida EL, Chan YK, Steel CJ. Dietary fibre sources in frozen part-baked bread: Influence on technological quality. LWT—Food Sci Technol 2013;53:262–70.
[6] Maphosa Y, Jideani VA. Dietary fiber extraction for human nutrition—a review. Food Rev Int 2016;32(1):98–115.
[7] Prakongpan T, Nitithamyong A, Luangpituksa P. Extraction and application of dietary fiber and cellulose from pineapple cores. J Food Sci 2002;67:1308–13.
[8] Dhingra D, Michael M, Rajput H, Patil RT. Dietary fibre in foods: a review. J Food Sci Technol 2012;49(3):255–66.
[9] Yamazaki E, Murakami K, Kurita O. Easy preparation of dietary fiber with the high-water holding capacity from food sources. Plant Foods Hum Nutr 2005;60:17–23.
[10] Yangilar F. The application of dietary fibre in food industry: structural features, effects on health and definition, obtaining and analysis of dietary fibre: a review. J Food Nutr Res 2013;3:13–23.
[11] McKee LH, Latner TA. Underutilized sources of dietary fiber: a review. Plant Foods Hum Nutr 2000;55:285–304.

[12] Figuerola F, Hurtado ML, Estevez AM, Chiffelle I, Asenjo F. Fibre concentrates from apple pomace and citrus peel as potential fibre sources for food enrichment. Food Chem 2005;91:395–401.

[13] Nelson AL. High-fibre ingredients. Eagan Press Handbook Series. St. Paul: Eagan Press; 2001.

[14] Lambo AL, Öste R, Nyman MEG-L. Dietary fibre in fermented oat and barley β-glucan rich concentrates. Food Chem 2005;89:283–93.

[15] Lafiandra D, Riccardi G, Shewry PR. Improving cereal grain carbohydrates. J Cereal Sci 2014;59:312–26.

[16] Saulnier L, Guillon F, Chateigner-Boutin A-L. Cell wall deposition and metabolism in wheat grain. J Cereal Sci 2012;56:91–108.

[17] Vardakou M, Palop CN, Christakopoulus P, Faulds CB, Gasson MA, Narbad A. Evaluation of the prebiotic properties of wheat arabinoxylan fractions and induction of hydrolase activity in gut microflora. Int J Food Microbiol 2008;123:166–70.

[18] Lu ZX, Walker KZ, Muir JG, Mascara T, O'Dea K. Arabinoxylan fiber, a byproduct of wheat flour processing, reduces the postprandial glucose response in normoglycemic subjects. Am J Clin Nutr 2000;71:1123–8.

[19] Johansson L, Tuomainen P, Ylinen M, Ekholm P, Virkii L. Structural analysis of water-soluble and—insoluble β-glucans of whole grain oats and barley. Carbohyd Polym 2004;58:267–74.

[20] Delaney B, Nicolosi RJ, Wilson TA, Carlson T, Frazer S, Zheng GH, et al. Beta-glucan fractions from barley and oats are similarly antiatherogenic in hypercholesterolemic Syrian golden hamsters. J Nutr 2003;133:468–75.

[21] Hallfrisch J, Schofield DJ, Behall KM. Physiological responses of men and women to barley and oat extracts (NutrimX). II. Comparison of glucose and insulin responses. Cereal Chem 2003;80:80–3.

[22] BEDCA. Spanish food composition database. Available from: www.bedca.net

[23] Dermibas A. β-Glucan and mineral nutrient contents of cereals grown in Turkey. Food Chem 2005;90:773–7.

[24] Barikmo I, Ouattara F, Oshaug A. Protein, carbohydrate and fibre cereals from Mali-how to fit the results in a food composition table and database. J Food Comp Anal 2004;17:291–300.

[25] Li BW, Andrews KW, Pehrsson PR. Individual sugars, soluble, and insoluble dietary fiber contents of 70 high consumption foods. J Food Comp Anal 2002;15:715–23.

[26] Lamothe LM, Srichuwong S, Reuhs BL, Hamaker BR. Quinoa (*Chenopodium quinoa* W.) and amaranth (*Amaranthus caudatus* L.) provide dietary fibres high in pectic substances and xyloglucans. Food Chem 2015;167:490–6.

[27] Liu Z, Cheng F, Zhang G. Grain phytic acid content in Japonica rice as affected by cultivar and environmental and its relation to protein content. Food Chem 2005;89:49–52.

[28] Sumczynski D, Bubelovà Z, Fišera M. Determination of chemical, insoluble dietary fibre, neutral-detergent fibre and in vitro digestibility in rice types commercialized in Czech markets. J Food Comp Anal 2015;40:8–13.

[29] Braaten Jt, Wood PJ, Scott FW, Wolynetz MS, Lowe MK, Bradley-White P, et al. Oat β-glucan reduces blood cholesterol in hipercholesterolemic subjects. Eur J Clin Nutr 1994;48:465–74.

[30] Brennan CS, Clearly LJ. The potential use of cereal (1→3,1→4)-β-glucans as functional food ingredients. J Cereal Sci 2005;42:1–13.

[31] Cordeiro L, Reinhardt V, Baggio C, Werner M, Burci L, Sasaki G, et al. Arabinan and arabinan-rich pectic polysaccaharides from quinoa (*Chenopodium quinoa)* seeds: structure and gastroprotective activity. Food Chem 2012;130(4):937–44.

[32] Dodevska MS, Djordjevic BI, Sobajic SS, Miletic ID. Characterisation of dietary fibre components in cereals and legumes used in Serbian diet. Food Chem 2013;141:1624–9.

[33] Østegard K, Björck I, Vainionpää J. Effects of extrusion cooking on starch and dietary fibre in barley. Food Chem 1989;34:215–27.
[34] Vasanthan T, Gaosong J, Li J. Dietary fiber profile of barley flour as affected by extrusion cooking. Food Chem 2002;77:35–40.
[35] Wang V-M, Klopfenstein CF. Effect of twin-screw extrusion on the nutritional quality of wheat, barley, and oat. Cereal Chem 1993;70(6):712–5.
[36] Varo P, Laine R, Koivistoinen P. Effect of heat treatment on dietary fibe: interlaboratory study. J AOAC 1983;66(4):933–8.
[37] Ralet M-C, Thibault J-F, Della Valle G. Influence of extrusion-cooking on the physico-chemical properties of wheat bran. J Cereal Sci 1990;11:249–59.
[38] Honc I, Sluková M, Vaculová K, Sedláčˇková I, Wege B, Fehling E. The effects of extrusion on the content and properties of dietary fibre components in various cultivars. J Cereal Sci 2016;68:132–9.
[39] Trinidad P, Trinidad TP, Mallillin AC, Loyola AS, Sagum RS, Encabo RR. The potential health benefits of legumes as a good source of dietary fibre. Br J Nutr 2010;103:569–74.
[40] Mallillin AC, Trinidad TP, Raterta R, Dagbay K, Loyola AS. Dietary fibre and fermentability characteristics of root crops and legumes. Br J Nutr 2008;100:485–8.
[41] Kutos T, Golob T, Kac M, Plestenjak A. Dietary fibre content of dry and processed beans. Food Chem 2003;80:231–5.
[42] Pérez-Hidalgo MA, Guerra-Hernández E, García-Villanova B. Dietary fibre in three raw legumes and processing effect on chickpeas by enzymatic-gravimetric method. J Food Comp Anal 1997;10:66–72.
[43] Aguilera Y, Martín-Cabrejas MA, Benítez V, Mollá E, López-Andréu FJ, Esteban RM. Changes in carbohydrate fraction during dehydration process of common legumes. J Food Comp Anal 2009;22(7–8):678–83.
[44] Almeida-Costa GE, Silva K, Pissini SM, Costa A. Chemical composition, dietary fibre and resistant starch contents of raw and cooked pea, common bean, chickpea and lentil legumes. Food Chem 2006;94:327–30.
[45] Dueñas M, Sarmento T, Aguilera Y, Benitez V, Mollá E, Esteban RM, et al. Impact of cooking and germination on phenolic composition and dietary fibre fractions in dark beans (*Phaseolus vulgaris* L.) and lentils (*Lens culinaris* L.). LWT—Food Sci Technol 2016;66:72–8.
[46] Martín-Cabrejas MA, Sanfiz B, Vida A, Mollá E, Esteban E, López-Andreu FJ. Effect of fermentation and autoclaving on dietary fiber fractions and antinutritional factors of beans (*Phaseolus vulgaris* L.). J Agric Food Chem 2004;52(2):261–6.
[47] Donangelo CM, Trugo LC, Trugo NMF, Eggum BO. Effect of germination of legume seeds on chemical composition and on protein and energy utilization in rats. Food Chem 1995;53:23–7.
[48] Martín-Cabrejas MA, Aguilera Y, Benítez V, Mollá E, López-Andréu FJ, Esteban RM. Effect of industrial dehydration on the soluble carbohydrates and dietary fibre fractions in legumes. J Agric Food Chem 2006;54:7652–7.
[49] Ghavidel A, Prakash J. The impact of germination and dehulling on nutrients, antinutrients, in vitro iron and calcium bioavailability and in vitro starch and protein digestibility of some legume seeds. LWT—Food Sci Technol 2007;40(7):1292–9.
[50] Martín-Cabrejas MA, Ariza N, Esteban RM, Mollá E, Waldron K, López-Andréu FJ. Effect of germination on the carbohydrate composition of the dietary fibre of peas (*Pisum sativum* L.). J Agric Food Chem 2003;51:1254–9.
[51] Alabi DA, Alausa AA. Evaluation of the mineral nutrients and organic food content of the seeds of *Lablab purpureus, Leucaena Leucocaphala* and *Mucuna utilis* for domestic consumption and industrial utilization. World J Agric Sci 2006;2:115–8.
[52] Martín-Cabrejas MA, Díaz MF, Aguilera Y, Benítez V, Mollá E, Esteban R. Influence of germination on the soluble carbohydrates and dietary fibre fractions in non-conventional legumes. Food Chem 2008;107(3):1045–52.

[53] Benítez V, Cantera S, Aguilera Y, Mollá E, Esteban RM, Díaz MF, et al. Impact of germination on starch, dietary fiber and physicochemical properties in non-conventional legumes. Food Res Int 2013;50(1):64–9.
[54] Sridhar KR, Seena S. Nutritional and antinutritional significance of four unconventional legumes of the genus *Canavalia*—a comparative study. Food Chem 2006;99(2):267–88.
[55] Grigelmo-Miguel N, Martin-Belloso O. Comparison of dietary fibre from by-products of processing fruits and greens and from cereals. LWT—Food Sci Technol 1999;32: 503–8.
[56] Rehinan Z, Rashid M, Shah WH. Insoluble dietary fibre components of food legumes as affected by soaking and cooking processes. Food Chem 2004;85:245–9.
[57] Marconi E, Ruggeri S, Cappelloni M, Leonardi D, Carnovale E. Physicochemical, nutritional, and microstructural characteristics of chickpeas (*Cicer arietinum* L.) and common beans (*Phaseolus vulgaris* L.) following microwave cooking. J Agric Food Chem 2000;48(12):5986–94.
[58] Peña MJ, Vergara CE, Carpita NC. The structures and architectures of plant cell walls define dietary fibre composition and the textures of foods. In: McCleary BV, Prosky L, editors. Advanced dietary fibre technology. Oxford: Blackwell Science Ltd; 2001. p. 42–60.
[59] Aldwairji MA, Chu J, Burley VJ, Orfila C. Analysis of dietary fibre of boiled and canned legumes commonly consumed in the United Kingdom. J Food Comp Anal 2014;36:111–6.
[60] Chang HL, Sang HO, Eun JY, Young SK. Effects of raw, cooked and germinated small black soybean powders on dietary fibre content and gastrointestinal functions. Food Sci Biotechnol 2006;15:635–8.
[61] Mataix J, García L, Mañas M, Martínez-Victoria E, Llopis J. Tabla de Composición de Alimentos. 4th ed. Granada: Universidad de Granada; 2003.
[62] Moreiras O, Carbajal A, Cabrera L, Cuadrado C. Tablas de Composición de los Alimentos". Madrid: Pirámide; 2001.
[63] Chau ChF, Huang YL. Comparison of the chemical composition and physicochemical properties of different fibers prepared from peel of *Citrus sinensis* L. cv. Liucheng. J Agric Food Chem 2003;51:2615–8.
[64] Rodriguez RAJ, Fernandez-Bolaños J, Guillen R, Heredia A. Dietary fibre from vegetable products as a source of functional ingredients. Trends Food Sci Technol 2006;17: 3–15.
[65] Cho SS, Prosky L. Application of complex carbohydrates to food product fat mimetics. In: Cho SS, Prosky L, Dreher M, editors. Complex carbohydrates in foods. New York: Marcel Dekker; 1999. p. 411–30.
[66] Larrauri JA, Goñi I, Martín-Carrón N, Rupérez P, Saura-Calixto F. Measurement of health-promoting properties in fruit dietary fibres: Antioxidant capacity, fermentability and glucose retardation index. J Sci Food Agric 1996;71:515–9.
[67] Saura-Calixto F. Antioxidant dietary fiber product: a new concept and a potential food ingredient. J Agric Food Chem 1998;46:4303–6.
[68] Grigelmo-Miguel N, Martín-Belloso O. Influence of fruit dietary fibre addition on physical and sensorial properties of strawberry jams. J Food Eng 1999;41:13–21.
[69] Chi-Fai Ch, Ya-Ling H, Mao-Hsiang L. In vitro hypoglycaemic effects of different insoluble fiber-riche fractions prepared from the peel of *Citrus sinensis* L. cv. Liucheng. J Agric Food Chem 2003;51:6623–6.
[70] Yokoyama WH, Hudson CA, Knuckles BE, Chiu MM, Sayre RN, Turnland JR, et al. Effect of barley β-glucan in durum wheat pasta on human glycaemic response. Cereal Chem 1997;74:293–6.
[71] Cavallero A, Empilli S, Brighenti F, Stanca AM. High (1/3) (1/4)-b-D-glucan fractions in bread making and their effect on human glycaemic response. J Cereal Sci 2002;36: 59–66.

[72] Symons LJ, Brennan CS. The physicochemical and nutritional evaluation of wheat breads supplemented with (1/3) (1/4)-b-D-glucan rich fractions from barley. J Food Sci 2004;69:463–7.
[73] Brennan CS, Tudorica CM, Kuri V. Soluble and insoluble dietary fibres (non-starch polysaccharides) and their effects on food structure and nutrition. Food Ind J 2002;5: 261–72.
[74] Konuklar G, Inglett GE, Warner K, Carriere CJ. Use of a betaglucan hydrocolloidal suspension in the manufacture of low-fat cheddar cheeses: textural properties by instrumental methods and sensory panels. Food Hydrocoll 2004;18:535–45.
[75] Russo M, Bonaccorsi I, Torre G, Sarò M, Dugo P, Mondello L. Underestimated sources of flavonoids, limonoids and dietary fibre: availability in lemon's by-products. J Funct foods 2014;9:18–26.
[76] Russo M, Bonaccorsi I, Inferrera V, Dugo P, Mondello L. Underestimated sources of flavonoids, limonoids and dietary fiber: availability in orange's by-products. J Funct 2015;12:150–7.
[77] Teixeria F, Rodrigues L, Sampaio B, Aline F, Bizzani M, Picolli L. Biological properties of apple pomace, orange bagasse and passion fruit peel as alternative sources of dietary fibre. Bioact Carbohyd Dietary Fibre 2015;6:1–6.
[78] Benítez V, Mollá E, Martín-Cabrejas MA, Aguilera Y, López-Andréu FJ, Esteban RM. Effect of sterilisation on dietary fibre and physicochemical properties of onion byproducts. Food Chem 2011;127:501–7.
[79] Benítez V, Mollá E, Martín-Cabrejas MA, Aguilera Y, López-Andréu FJ, Esteban RM. Onion (*Allium cepa* L) byproducts as source of dietary fiber: physicochemical properties and effects on serum lipid levels in high-fat fed rats. Eur Food Res Technol 2012;234:617–25.
[80] López-Marcos MC, Bailina C, Viuda-Martos M, Pérez-Alvarez JA, Fernández-López J. Properties of dietary fibers from agroindustrial coproducts as source for fiber-enriched foods. Food Bioprocess Technol 2015;8:2400–8.
[81] Amaya-Cruz DM, Rodríguez-González S, Pérez-Ramírez IF, Loarca-Piña G, Amaya-Llano S, Gallegos-Corona MA, et al. Juice by-products as a source of dietary fibre and antioxidants and their effect on hepatic steatosis. J Funct Foods 2015;17:93–102.
[82] Al-Sayed HMA, Ahmed AR. Utilization of watermelon rinds and sharlyn melon peels as a natural source of dietary fiber and antioxidants in cake. Ann Agric Sci 2013;58(1): 83–95.

CHAPTER 8

High Fiber Diet in the Clinical Setting

Rodney A. Samaan

It is hard to believe that I graduated from medical school without a single lecture on nutrition. I went on to complete a 4-year residency in internal medicine and pediatrics with some very minor discussion on nutrition during my pediatric training. The extent of the latter discussion was on how to order total parenteral nutrition (TPN), which was often given to many children with inflammatory bowel disease and other conditions of the digestive system (i.e., necrotizing enterocolitis). I then entered a 3-year cardiology fellowship program, and although I discussed lipid management with my colleagues, there was never one comprehensive discussion about nutrition. When I started practicing cardiology, I began to look more closely at my diet, as I had always been healthy and active. I quickly realized that despite all the education, many physicians were not aware of the nutritional contribution to heart disease. Many physicians are ill equipped to counsel patients on the nutritional benefits of fiber in preventing heart disease. I found that my patients also were not informed on what the appropriate diet they should be following. Hence, the origins of this book, which came out of a necessity to understand the problem for myself.

Most nutritional guidelines recommend at least 25–30 g of fiber a day, which includes some fruits and vegetables. There are a number of popular books, videos, and documentaries that demonstrate the detrimental effects of poor nutritional habits on peoples' health. There were many influences to my nutritional education including a book by Robert Lustig (such as *Fat Chance*), documentaries such as *Forks over Knives and Food Inc.*, the famous video by Lustig titled *Sugar: the Bitter Truth*, and finally the most recent documentary called: *Fed Up* (narrated by Katie Couric). These books and videos made me realize that not only do we need to decrease out sugar intake, but also pay attention to how much fiber we eat in our diets. Furthermore, this information needs to be explained and disseminated to cardiologist,

Dietary Fiber for the Prevention of Cardiovascular Disease
http://dx.doi.org/10.1016/B978-0-12-805130-6.00008-2

internist, family doctors, nutritionist, medical students, midlevel providers, nurses, and medical assistants

I hope that in reading the prior chapters on fiber's prevention of obesity and cardiovascular disease, providers and researchers will better understand how both soluble and insoluble can lead to much lower rates of heart disease, diabetes, and obesity. We have come a long way since I was in medical school and the same for any physician who graduated more than 10 years ago. According to gastroenterologist, Michael Gershon quote in his book: *The Second Brain:* "When I was at medical school, I was taught erroneously that the brain controlled everything—including the gut, In fact, if you cut the vagus nerve—the major nerve between the brain and the gut—the gut would soldier on. We now know it can work completely independently of the brain and spinal cord. While the 'first brain' gets on with religion, philosophy, and poetry, the 'second brain' deals with the messy business of digestion" [1]. He also estimates that the GI tract is "... home to up to 100 million neurons—as many as the spinal cord—and about 40 neurotransmitters—as many as we have in the brain. About 90 to 95 percent of the vagus fibers are carrying signals from the gut to the brain—not the other way round" [1].

We are learning a great deal about the digestive system and fiber is in many ways the oil that keeps the engine running smoothly. Without fiber, we are doomed to many chronic illnesses including: diabetes, heart disease, gallstones, constipation, and colon cancer. As research advances, more and more data is entering the public and scientific world that continues to shed more light on this fascinating aspect of the digestive system. However, this book is only one small contribution to a much more complex field and as such, only one small piece of the puzzle. Nutrition has always been a complex and controversial subject that is not taught in medical schools, but this is also gradually changing.

For many patients, there is a significant amount of misinformation and confusion about food/diets that it is no surprise that many patients are unsure about how to eat a healthy diet. Even the food industry can somehow label foods as heart healthy (i.e., Cheerios has very little fiber content, but does contain whole grains which are processed) based on inadequate research; yet still Cheerios is able to gain a "heart healthy" label on the front of the box. Many patients are shocked when I tell them that I would rather eat an apple a day compared to a banana every day. All of the patients ask me the same question: Where will I get my potassium? I always respond that many foods have potassium and the key is to eat fruits that contain a balance of sugar and fiber.

About 1-year ago, I printed a list of high fiber foods and low fiber foods and placed them into two columns: one column for foods to eat (high fiber) and one column for foods to avoid (low fiber foods). My patients were skeptical, but I always told them to try it for 1 week and to see how you feel on this diet. I also emphasized the need for plenty of water consumption and exercise.

More than a 100 years ago, people in Britain were eating close to 100 g of fiber. Even earlier in our history, some claim, "our ancestors ate over 150 grams of fiber per day" [2]. Currently, most people consume about 16 g of fiber per day [2]. In one recent survey, about 97% of people in the USA do not consume the daily recommend fiber intake [2]. The recommendations are now to eat about 30 g of fiber per day.

All the seven chapters explain in detail the incredible health benefits of a high-fiber diet and how fiber can reduce the risk of heart disease, colon cancer, diabetes, and obesity. Based on my patient experience, in patients who are compliant with this diet, I have seen the recent risk reductions come to fruition in a live clinical setting over a 6-month period, in which they participated in a diabetes and nutrition education class (Table 8.1).

Some of these examples include the following:

1. Increased energy
2. Better mood
3. Reduction in HbA1c from 13 to 7 (in 3 months) (Table 8.1)
4. Weight loss of up to 20 lbs
5. Cholesterol reductions
6. Improved blood pressure

By prescribing a high fiber diet to all my patients and providing a general nutrition counseling instead of only drugs, I have hencegained their trust and patients also appreciate it when doctors do more than order tests and prescribe new medications.

A high fiber diet is a very simple diet that requires you to eat more than 30 g of fiber (with plenty of water). This book clearly demonstrates how both insoluble and soluble fiber help change the gut microbial environment, nourish the colonic bacteria, which then produce short-chain fatty acids and butyrate as well as many other complex compounds that help to reduce inflammation and hence decrease insulin secretion and decrease glucose instability.

Table 8.1 Patients on a High Fiber Diet achieved on average a 1% reduction in HbA1c.

Starting HbA1C	Ending HbA1c	Change	Change (%)
11.9	12.6	−0.7	5.88
11.6	8.3	3.3	28.45
7.4	7.5	−0.1	1.35
5.9	6.0	−0.1	1.69
11.8	11.1	0.7	5.93
11.1	8.9	2.2	19.82
13.3	12.5	0.8	6.02
12.4	13.6	−1.2	9.68
11.7	10.4	1.3	11.11
13.0	8.1	4.9	37.69
13.0	12.8	0.2	1.54
10.7	9.4	1.3	12.15
14.5	10.0	4.5	31.03
9.2	9.4	−0.2	2.17
10.0	12.1	−2.1	21.00
14.4	8.3	6.1	42.36
5.6	5.5	0.1	1.79
9.7	10.6	−0.9	9.28
10.5	8.8	1.7	16.19
10.1	8.4	1.7	16.83
12.9	11.3	1.6	12.40
12.3	8.1	4.2	34.15
6.2	6.1	0.1	1.61
14.0	13.3	0.7	5.00
11.5	11.0	0.5	4.35
11.1	10.0	1.1	9.91
10.1	10.7	−0.6	5.94
12.0	13.7	−1.7	14.17
12.9	10.7	2.2	17.05
11.1	10.3	0.8	7.21
13.7	7.7	6.0	43.80
9.9	8.4	1.5	15.15
12.7	10.7	2.0	15.75
13.0	10.1	2.9	22.31
14.2	9.6	4.6	32.39
9.3	9.9	−0.6	6.45
		1.36	10.40

In conclusion, doctors should recommend a high fiber diet to all our patients; there is no doubt that this diet can reduce this country's diabetes and colon cancer burden. We need to consider the following facts from different cultural and ethnic groups in regards to the consumption of sugar, fat, starch, protein, salt, and fiber intake. According to the World Health Organization and Dr. Darwin Labarthe, in the most extreme and ideal situation, the peasant agriculturist consumed 10%–15% fat, 5% sugar, 60%–75% starch, 10%–15% protein, with a total of 60–120 g of fiber per day [3,4]. In the modern world, we are consuming 40% of our calories from fat, 20% from sugar, 25%–30% starch, 12% protein, and a total of 15–20 g of fiber per day [3,4]. We need to go back to our roots and reduce our fat and sugar content and increase our starch and fiber intake.

REFERENCES

[1] Gershon M. The second brain: a groundbreaking new understanding of nervous disorders of the stomach and intestine. New York: HarperCollins; 1999. 336p.
[2] Sifferlin A, Oaklander M, Park A. Here are your new food rules. Time Magazine 2015;186(27).
[3] World Health Organization. Diet, Nutrition, and the prevention of Chronic Diseases; Techinal Report Series 797. Wold Health Organization Study Group, Geneva, Switzerland; 1990. 102p.
[4] Labarthe D. Epidemiology and prevention of cardiovascular diseases: a global challenge. 2nd ed. Sudbury: Jones & Bartlett Learning; 2011. 709p.

INDEX

D

E

F

G

Q

R

S

Printed in the United States
By Bookmasters